Modern Acoustics and Signal Processing

Series Preface for Modern Acoustics and Signal Processing

In the popular mind, the term "acoustics" refers to the properties of a room or other environment—the acoustics of a room are good or the acoustics are bad. But as understood in the professional acoustical societies of the world, such as the highly influential Acoustical Society of America, the concept of acoustics is much broader. Of course, it is concerned with the acoustical properties of concert halls, classrooms, offices, and factories—a topic generally known as architectural acoustics, but it also is concerned with vibrations and waves too high or too low to be audible. Acousticians employ ultrasound in probing the properties of materials, or in medicine for imaging, diagnosis, therapy, and surgery. Acoustics includes infrasound—the wind driven motions of skyscrapers, the vibrations of the earth, and the macroscopic dynamics of the sun.

Acoustics studies the interaction of waves with structures, from the detection of submarines in the sea to the buffeting of spacecraft. The scope of acoustics ranges from the electronic recording of rock and roll and the control of noise in our environments to the inhomogeneous distribution of matter in the cosmos.

Acoustics extends to the production and reception of speech and to the songs of humans and animals. It is in music, from the generation of sounds by musical instruments to the emotional response of listeners. Along this path, acoustics encounters the complex processing in the auditory nervous system, its anatomy, genetics, and physiology—perception and behavior of living things.

Acoustics is a practical science, and modern acoustics is so tightly coupled to digital signal processing that the two fields have become inseparable. Signal processing is not only an indispensable tool for synthesis and analysis, it informs many of our most fundamental models for how acoustical communication systems work.

Given the importance of acoustics to modern science, industry, and human welfare Springer presents this series of scientific literature, entitled Modern Acoustics and Signal Processing. This series of monographs and reference books is intended to cover all areas of today's acoustics as an interdisciplinary field. We expect that scientists, engineers, and graduate students will find the books in this series useful in their research, teaching and studies.

William M. Hartmann
Series Editor-in-Chief

More information about this series at http://www.springer.com/series/3754

Robert C. Maher

Principles of Forensic Audio Analysis

Springer

Robert C. Maher
Department of Electrical and Computer Engineering
Montana State University
Bozeman, MT, USA

ISSN 2364-4915 ISSN 2364-4923 (electronic)
Modern Acoustics and Signal Processing
ISBN 978-3-319-99452-9 ISBN 978-3-319-99453-6 (eBook)
https://doi.org/10.1007/978-3-319-99453-6

Library of Congress Control Number: 2018957136

This Springer imprint is published by the registered company Springer Nature Switzerland AG
The registered company address is: Gewerbestrasse 11, 6330 Cham, Switzerland

To my parents, Jane Crawford Maher,
and the late Prof. Louis J. Maher, Jr.
(1933–2018).

Preface

Audio forensics is an essential specialty in modern forensic science. This book provides the fundamental background necessary to understand and participate in this exciting and important field of study. Modern audio forensic analysis combines skills in digital signal processing, the physics of sound propagation, acoustical phonetics, audio engineering, and many other fields.

Scientists and engineers who work in the field of audio forensics are called upon to address issues of authenticity, quality enhancement, and signal interpretation for audio evidence that is important to a criminal law enforcement investigation, an accident investigation board, or an official civil inquiry of some kind.

Expertise in audio forensics has never been more important. Common recordings from emergency call centers and police radio dispatch continue to be important. Yet today it is the fact that inexpensive portable audio/video recording systems are now in such widespread use that forensic evidence from the scene of a civil or criminal incident increasingly involves dashboard recorders in police cars, vest-pocket personal recorders worn by law enforcement officers, smartphone recordings from bystanders, and security surveillance systems in public areas and businesses. These increasingly ubiquitous audio recording devices will undoubtedly increase the quantity and quality of audio forensic material available for many investigations.

Utilizing new research findings and both historical and contemporary casework examples, this book blends audio forensic theory and practice in a manner intended for any scientifically literate reader. Extensive examples, supplementary material, and bibliographic references are also included for those who are interested in delving deeper into the field.

Bozeman, MT, USA — Robert C. Maher

Acknowledgments

I wish to acknowledge the contributions of many professional colleagues in the field of audio forensics who have generously shared their knowledge and expertise with me over the last decade. In particular, the Audio Engineering Society's Technical Committee on Audio Forensics, originally founded by the late Roy Pritts and the late Richard Sanders, has provided an excellent set of international conferences and workshops that have inspired much of my work. I am pleased to thank AES members Durand Begault, Eddy Brixen, Catalin Grigoras, Bruce Koenig, Douglas Lacey, and Jeff Smith.

I gratefully acknowledge fruitful conversations and professional interaction with Steven Beck, Angelo Borzino, Jack Freytag, Kurt Fristrup, Tom Owen, Steven R. Shaw, and Gregg Stutchman. Douglas Keefe provided clear and concise guidance related to auditory anatomy and physiology, and Luke Haag patiently improved my understanding of firearms. I am also appreciative of many of my recent students who have worked on research projects in areas related to audio forensic analysis, including Cameron Blegen, Matthew Blunt, Ethan Hoerr, Darrin Reed, Tushar Routh, and Joseph Studniarz.

My academic research work in gunshot acoustical forensics is supported by grants from the United States Department of Justice, National Institute of Justice. I thank Monte Evans and Gregory Dutton from the NIJ Office of Justice Programs for their assistance.

Many of the figures in this book were created from images obtained using commercial software, including *Adobe® Audition®* and *MathWorks® MATLAB®*.

I joyfully acknowledge the outstanding enthusiasm, editorial assistance, and patience of the team at Springer. Thank you in particular to Sara Kate Heukerott, Sanaa Ali-Virani, and William M. Hartmann. You made this book possible.

I am incredibly fortunate to have the love and support of my wife, Lynn, who for more than thirty years has tolerated my many hours spent at the office in the evening and the weekends scheming and tinkering; and our sons, Maxwell and Henry, who have grown into remarkable young men whom we cherish and admire. Thank you.

Contents

About the Author

Robert C. Maher is a professor of Electrical and Computer Engineering at Montana State University, Bozeman. An experienced professional engineer, educator, and entrepreneur, his research and teaching interests are in the field of digital signal processing, with particular emphasis in digital audio, audio forensic analysis, digital music synthesis, and acoustics. He has written and lectured extensively in the field of audio engineering and also serves as an expert witness in civil and criminal cases.

Dr. Maher is a fellow of the Audio Engineering Society, senior member of IEEE, associate member of the American Academy of Forensic Sciences, and a member of the Acoustical Society of America. He lives in Bozeman, Montana, where he enjoys spending his spare time as an amateur musician and trail runner.

Chapter 1
Introduction to Forensic Audio Analysis: Authenticity, Enhancement, and Interpretation

Audio forensics is a branch of the broad field of forensic science. Forensic science generally refers to evaluation of evidence that may ultimately be used in court or as part of some other formal investigation. Audio forensics, therefore, refers to the acquisition, analysis, and interpretation of audio recordings as part of an official investigation, such as in preparation for a civil or criminal trial, or as part of the investigation of an accident or some other incident involving audio evidence (Maher 2009, 2015).

What sort of questions arises in audio forensic investigations? Typical investigations involve one or more of three primary concerns: authenticity, enhancement, and interpretation.

Authenticity is important in forensic investigations because the significant conclusions the investigator draws from the audio recording depend upon the circumstances under which the recording was made. If it turns out that the recording was altered deliberately—or inadvertently—prior to the investigation, the entire examination is called into question. Similarly, if there is a deliberate or inadvertent mistake about the place and/or time that the recording occurred, the examination will be irrelevant. Audio forensic examiners must assess the chain of custody of the evidence, take steps to uncover deliberate tampering, and provide safeguards to protect against inadvertent alteration.

Cases involving audio forensic evidence often involve requests for *audio enhancement*. Many audio recordings that end up being of forensic interest take place under non-ideal acoustic circumstances: poor microphone position, strong or fluctuating background noise, the talkers may not enunciate clearly, the signal of interest is weak, etc. In these circumstances, the audio information of interest to the investigation must be processed to emphasize the features of interest. Enhancement may be particularly important when audio forensic evidence will be presented in court, because judges and juries generally do not have experience listening to and interpreting noisy audio nor do they have the time to listen to the material multiple times at different playback levels. Audio presentations in court seldom involve ideal

R. C. Maher, *Principles of Forensic Audio Analysis*, Modern Acoustics and Signal Processing, https://doi.org/10.1007/978-3-319-99453-6_1

playback conditions, so it is vital to choose an appropriate degree of enhancement for the circumstances.

Interpretation of audio evidence may involve many questions of forensic interest, such as reconstructing timelines, transcribing dialog, and identifying unknown sounds. Questions addressed by audio forensic examination are often based on an investigator's theory about the circumstances of a crime, or in the context of other physical evidence and witness accounts.

Audio recordings provide several potential advantages for an investigation compared to film, video, and eyewitness observations, such as the ability to collect information from all directions, not just in a particular field of view. Audio recordings provide a sequential time record of events as an objective observation, rather than as a witness' subjective recollection.

Audio recordings may also have obvious shortcomings in an investigation, such as the general difficulty in determining direction and orientation of the sound source with respect to the recording microphone if only a single, monophonic recording is available. Other shortcomings may include the limited dynamic range of the recording: very subtle and quiet sounds may not appear with sufficient resolution in the recording, while very loud sounds may be "clipped" if they exceed the maximum limit of the recording system. Perhaps the most common issue with forensic audio recordings is the presence of interfering noise or extraneous sounds that can obscure the low-level sounds of interest to the investigation. Recordings that are of very high quality and intelligibility are often used directly by the individuals performing the investigation without being presented to an audio forensic examiner.

The chapters of this book follow a progression beginning with a few basic principles of acoustics and psychoacoustics, the history of audio forensics, and the common procedures of an audio forensic examination. The later chapters deal with authenticity, enhancement, and interpretation of audio evidence. Finally, the book concludes with an overview of expert reports and testimony and consideration of a few specific topics of current interest in this field.

References

Maher, R. C. (2009). Audio forensic examination: Authenticity, enhancement, and interpretation. *IEEE Signal Processing Magazine, **26***, 84–94.

Maher, R. C. (2015). Lending an ear in the courtroom: Forensic acoustics. *Acoustics Today, **11***(3), 22–29.

Chapter 2
Fundamentals of Audio Signals and Systems

The study of audio forensics is based on the foundation of acoustics and audio engineering. There are shelves of books and entire college-level courses devoted to the details of acoustics, so the following description of several important principles barely scratches the surface of the underlying physics and engineering. Nonetheless, it is important to have an introduction to the terminology and key features in order to appreciate the fascinating field of acoustics as applied to audio forensics.

Audio forensic evidence generally comprises an *audio recording*. The recording is an abstract representation of sounds detected in the air by means of a microphone, converted into electricity, and then stored in some sort of fixed medium, such as magnetic tape, magnetic or optical disc, or semiconductor memory. Audio recordings may be analog or digital, which refers to the representation of the audio information within the recording and playback systems.

The study of *acoustics* involves the physical principles of sound propagation in the air. In order to understand and interpret forensic recordings, it is important to have an understanding of acoustical concepts so that sounds detected in the recording can be analyzed and attributed to the known characteristics of sound reflection, absorption, diffraction, and reverberation.

2.1 Sound

Sound in the air is the result of vibration. Unlike wind, for which the air particles move steadily over some substantial distance, a vibrating surface causes the particles to move back and forth a short distance as the surface moves. As the vibrating surface moves outward, the air particles near the surface get *pushed* (compressed), while during the other half of the vibration cycle, the surface moves inward, and the air particles near the surface get *pulled* (expanded or rarified). The alternating compression and expansion adjacent to the vibrating surface exerts a corresponding push and pull on the air particles next to them, which in turn push-pull the air

R. C. Maher, *Principles of Forensic Audio Analysis*, Modern Acoustics and Signal Processing, https://doi.org/10.1007/978-3-319-99453-6_2

particles in the next layer of air and so forth, causing a propagating *wave* of alternating high- and low-pressure regions. The alternating high- and low-pressure wave fronts propagate away from the vibrating surface, as shown in Fig. 2.1.

The sound wave is a *longitudinal* disturbance, meaning that the air particles move back and forth from their equilibrium position in the direction the wave front is moving. This longitudinal motion is difficult to sketch graphically, so the graphical depiction of sound is typically a two-dimensional graph of *acoustic pressure* vs. time, which tends to cause the misconception that the sound wave is somehow moving "up and down" as it propagates. A better understanding is that the particles move "in and out" (and the corresponding acoustic pressure increases and decreases) as the wave passes through (Kinsler et al. 2000).

It is important to realize that the acoustic pressure is a *very small fluctuation* compared to normal atmospheric pressure. The earth's gravity holds the approximately 100 km layer of our atmosphere against the earth, resulting in the nominal sea level air pressure, 1 at = 101.325 kPa = $\sim 1 \times 10^5$ Pa. Typical pressure fluctuations due to sound vibrations are tiny in comparison: in the range of millipascals ($\sim 10^{-3}$ Pa). In fact, the quietest audible sounds have a pressure amplitude of approximately 20 μPa (2×10^{-5} Pa), or just *1 part in 5 billion* compared to the nominal atmospheric pressure.

Very loud sounds like at a rock-and-roll concert or near an industrial machine may have an amplitude of 1 Pa or more, and despite being dangerously loud to our ears, this is still just 1 part in 50,000 compared to the nominal atmospheric pressure.

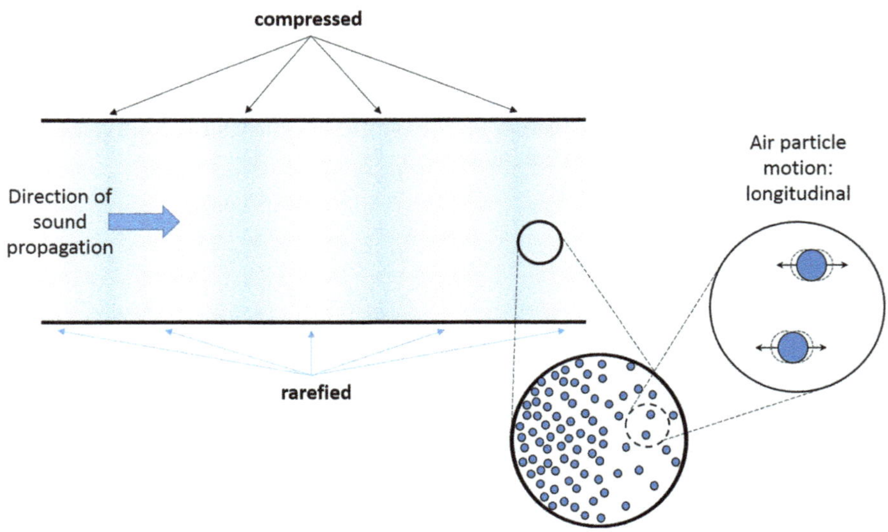

Fig. 2.1 Air particle motion in a sound wave: longitudinal (forward and backward) parallel to the direction of propagation

2.2 Sound Pressure Level

The audible acoustic pressure range from 2×10^{-5} Pa to 1 Pa results in numbers that are rather unwieldy for note-taking and printing. It has become customary to specify the audible range of sound pressure in a logarithmic fashion so that the quietest audible sound has a level of zero and the loudest commonly encountered sound has a level represented with just two or three digits. This scientific representation uses the *bel* [B], which is the base-10 logarithm of the ratio of two quantities expressing *power* in watts [W] or expressing *intensity* [W/m²].

$\text{bel} = \log_{10}(\text{power}_1/\text{power}_0)$

or

$\text{bel} = \log_{10}(\text{intensity}_1/\text{intensity}_0)$

In order to use the bel definition for sound, there needs to be a conversion from acoustic pressure [pascal] to acoustic intensity [watts/m²]. The relationship for sound waves is that acoustic intensity is proportional to the square of acoustic pressure. This allows the bel to be defined in terms of pressure as:

$\text{bel} = \log_{10}(\text{pressure}_1^2/\text{pressure}_0^2) = 2 \log_{10}(\text{pressure}_1/\text{pressure}_0)$

The customary representation uses the *decibel* [dB], where the *deci* prefix indicates that the precision is 1/10 of a bel. There are 10 dB in 1 bel, so a measurement in dB is 10 times the measurement expressed in bel.

$\text{decibel [dB]} = 20 \log_{10}(\text{pressure}_1/\text{pressure}_0)$

The *sound pressure level* (SPL) expressed in decibels uses the definition that pressure_0 is 20 μPa (20 μPa = 0.00002 Pa), and pressure_1 is the *effective pressure* (root-mean-square or RMS) measured with a microphone (Kinsler et al. 2000). The choice of 20 μPa as the reference pressure is appropriate because it roughly corresponds to the human threshold of hearing, meaning that an acoustic signal with effective pressure of 20 μPa corresponds to zero dB SPL. An acoustic signal at 100 dB SPL is very loud and roughly corresponds to a sound level that is painful to the ear. Thus, the range of practical sound levels for human audition is 0–100 dB SPL. Measurements with a sound level meter should always include the dB label and sound pressure reference, e.g., “60 dB SPL re 20 μPa” (Hartmann 2013).

Because the human ear has nonuniform sensitivity as a function of frequency, sound pressure level measurements are often made using a *weighting filter* that approximates the frequency dependence of the ear’s sensitivity (see Section 2.5.3 below).

The sound wave—the pressure disturbance of alternating high and low pressure—progresses through the air at a rate referred to as the *speed of sound*. The speed of sound depends upon the relationship between the acoustic pressure and the resulting vibratory motion (particle velocity) of the air particles. At 20 °C (room temperature), the speed of sound in air is 343 m/s. In comparison, the speed of light is 3×10^8 m/s, or about a million times faster than sound.

Convenient rules of thumb for sound propagation include the American approximations that sound travels about 1 foot per millisecond and requires about 5 s to travel a mile. Metric rules of thumb include about 35 cm per millisecond and about 3 s to travel a kilometer.

2.3 Wavelength, Frequency, and Spectrum

If the sound source is oscillating, such as a loudspeaker cone moving in and out or a guitar string vibrating back and forth, the sound will consist of alternating high-pressure and low-pressure cycles. The amount of time required for one oscillation is known as the *period* of the vibration. For example, the period of a vibrating string is the amount of time required for the string to move from one extreme to the other extreme and then back to the original position, completing 1 *cycle* of the oscillation. During the time it takes for one oscillation (one period), the resulting sound pressure disturbance will move away through the air at the speed of sound a certain distance, known as the *wavelength*, expressed in [meters/cycle]. In other words, the *wavelength* is the distance traveled by the sound wave in the time of one *period* of the oscillation.

Sound oscillations are commonly expressed as an oscillation *rate*: how many cycles of the oscillation occur in 1 s [cycles/second]. The oscillation rate is the *frequency* of the oscillation. It is customary to use the unit *hertz* (abbreviated Hz) for a cycles/second frequency measurement.

If the oscillation is less frequent (*low frequency*), the period of each cycle will be greater in duration, and therefore the wave will travel farther between cycles: lower frequency means longer wavelength. Conversely, if the oscillation is very rapid (high frequency), the pressure disturbance has little time to propagate between cycles: high frequency means shorter wavelength.

Mathematically, the relationship between frequency [f, cycles per second, or Hz] and wavelength [λ, meters] is $c = f\lambda$, where c is the speed of sound [meters/second]. Again, high-frequency sounds have short wavelengths, while low-frequency sounds have long wavelengths (Fig. 2.2).

The simplest type of sustained sound consists of energy at a single frequency. This is referred to as being a pure tone. The waveform for this single-frequency sound can be depicted graphically, as shown in Fig. 2.3, and is known mathematically as a *sine wave*, or as a *sinusoid*.

The vertical axis of the graph in Fig. 2.3 have a parameter such as pressure, voltage, or displacement, and the horizontal axis represents time. The figure shows a single cycle of the sine wave, which in this example takes 1 ms (1/1000 of a second). Since one cycle has a *period* $T = 1$ ms, the *frequency* of this waveform ($1/T$) is 1 kHz (1000 cycles per second). If observed for a longer time interval, a sustained 1 kHz sinusoid looks like Fig. 2.4.

For example, the sound of a person whistling at a constant pitch, which is very nearly a pure tone, would cause the output voltage from a microphone to be a waveform for which each cycle appears like the figures above.

Because a pure tone (sinusoidal waveform) has energy only at the frequency of its repetition rate, a graph of the *spectrum* of a sinusoid looks like a single "frequency line." Theoretically, the spectrum of a 1 kHz sinusoid is as shown in Fig. 2.5: the spectrum contains no energy at any frequency except 1 kHz.

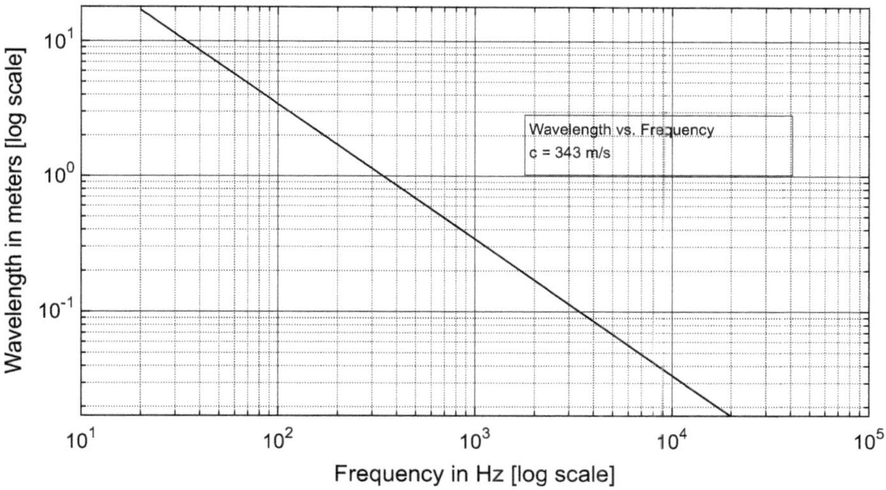

Fig. 2.2 The product of frequency (cycles per second) and wavelength (meters per cycle) is the speed of sound (meters per second)

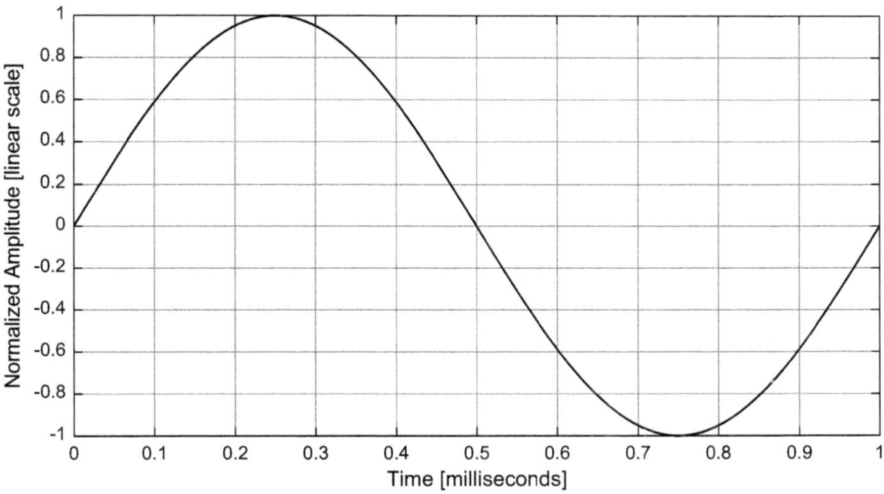

Fig. 2.3 A graph showing a single cycle of a 1 kHz sinusoidal wave, also known as a pure tone, consisting of sound energy at a single frequency

Sustained sounds that have a *periodic* (repetitive) waveform, like the sound of a person singing the vowel sound "ahhh" with constant pitch, have a more complicated spectrum than the single-frequency pure tone (Hartmann 2013). Periodic waveforms have a spectrum comprising *harmonics*, which means energy is present only at frequencies that are integer multiples of a *fundamental* frequency, denoted F_0. For example, a periodic waveform with fundamental frequency $F_0 = 1$ kHz is shown in Fig. 2.6.

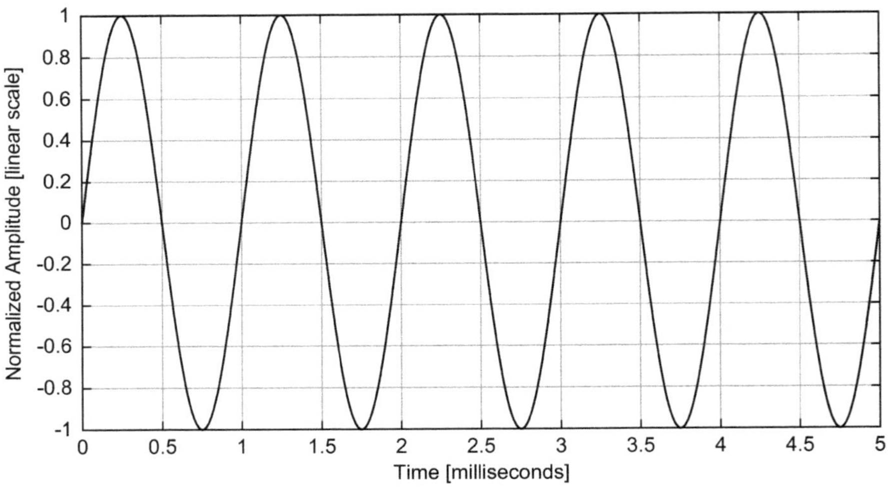

Fig. 2.4 A graph showing five cycles of a 1 kHz sinusoid

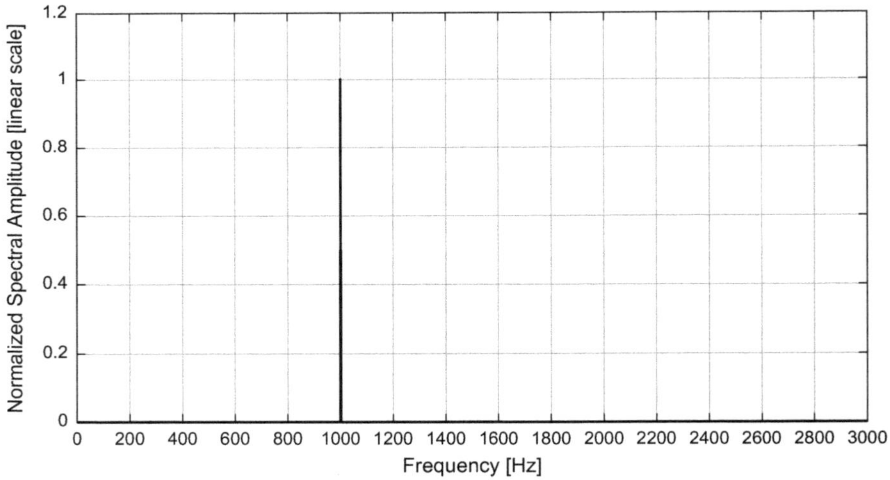

Fig. 2.5 The frequency spectrum magnitude of a 1 kHz pure tone. The spectrum of a pure sinusoid contains energy only at the single frequency of that tone

The magnitude spectrum corresponding to the periodic waveform of Fig. 2.6 is shown in Fig. 2.7. Note that the spectral components are all harmonics: *integer multiples* of the 1 kHz fundamental frequency (1 kHz, 2 kHz, 3 kHz, 4 kHz, etc.).

Given a purely periodic waveform like Fig. 2.6, it is possible to calculate numerically the harmonic amplitudes shown in Fig. 2.7 using a mathematical procedure known as the *Fourier series*. It is also possible to calculate the spectrum of a nonperiodic waveform using a mathematical procedure known as the *Fourier transform*. The mathematical justification and details are beyond the scope of this book, but we

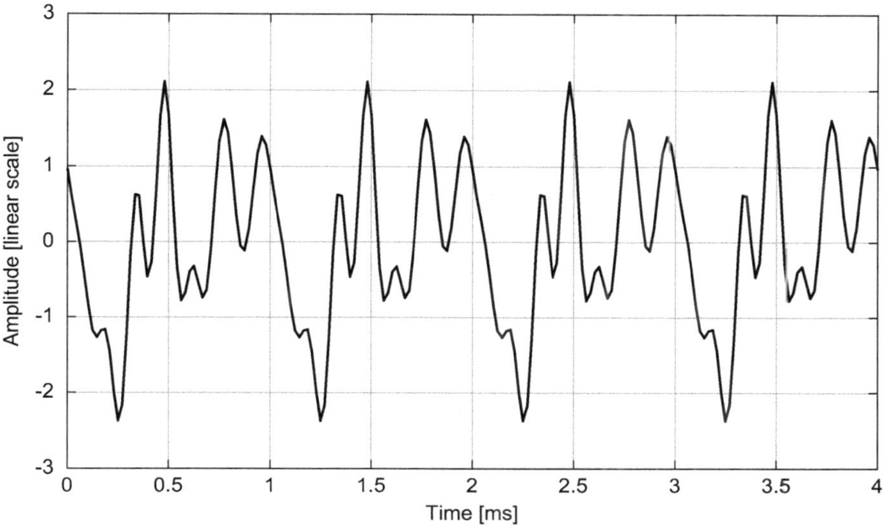

Fig. 2.6 Example periodic waveform with fundamental frequency 1 kHz

will use the Fourier transform to help understand the spectral characteristics of a variety of signals of interest in audio forensics (Hartmann 2013).

For example, a short segment of a recording of male speech is shown in Fig. 2.8. Note that the waveform is quasiperiodic in appearance, but each cycle is not a perfect replica of the others. The Fourier transform magnitude of the waveform is depicted in Fig. 2.9. Unlike the perfectly harmonic series of mono-frequency spikes in the spectrum of the theoretically infinitely long waveform of Fig. 2.6 and corresponding spectrum in Fig. 2.7, the Fourier analysis of a finite-length quasiperiodic speech waveform shows some broadening of the spectral lines and inharmonicity. This is due to cycle-to-cycle variation of the speech waveform and the finite length of time over which the signal is observed in the Fourier transform (Allen and Rabiner 1977).

If more than one sound source is present, the observed spectrum (Fourier transform magnitude) will contain an additive mixture of the spectral components from the various sources. This mixture happens deliberately when musical instruments play as an ensemble. Common-practice music of European heritage generally uses *consonant* combinations of musical frequencies so that the harmonics of the various fundamental frequencies will tend to overlap. For example, a musical signal with fundamental frequency 100 Hz will have harmonic energy in its spectrum at 200, 300, 400, 500, 600, 700, 800, 900, etc. Hz. If a simultaneous musical signal with fundamental frequency 150 Hz is present, that signal will have spectral energy at 150, 300, 450, 600, 750, 900, etc. Hz, which means that every other frequency component of the 150 Hz tone will "line up" with a harmonic of the 100 Hz tone (300, 600, 900, etc. Hz). In music theory, the relationship between a 100 Hz tone and a 150 Hz tone is referred to as a "perfect fifth" for reasons that are beyond the scope of this book, but suffice it to say that the harmonic frequency relationships play an important role in most kinds of music throughout the world.

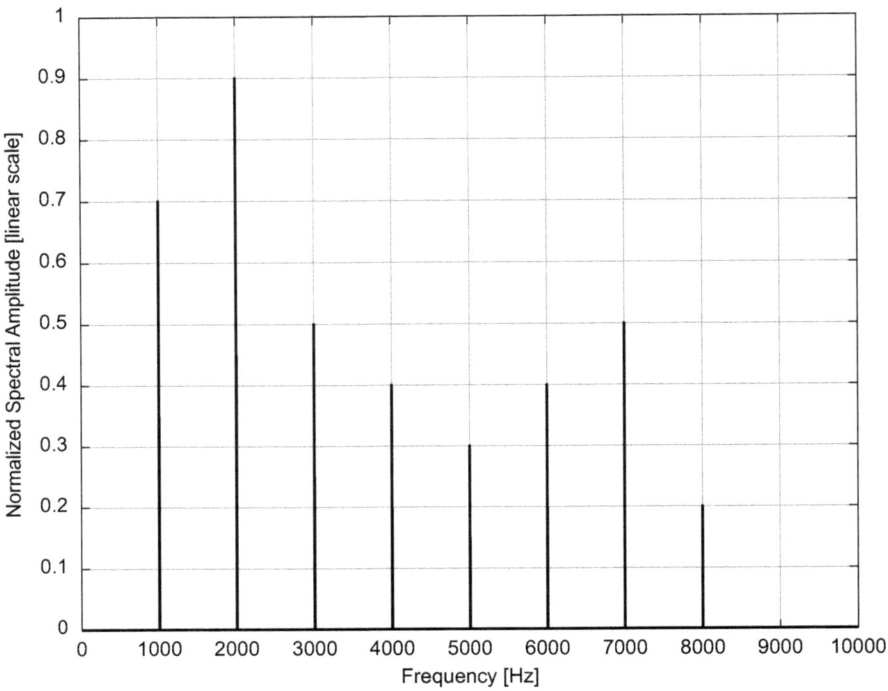

Fig. 2.7 Frequency spectrum of the perfectly periodic waveform of Fig. 2.6. The spectrum of a periodic waveform contains only harmonics (integer multiples) of a fundamental frequency. In this example, the fundamental is 1000 Hz, and there happen to be seven additional harmonic components. The harmonic amplitudes depend upon the details of the periodic waveform

For the purposes of audio forensic analysis, the *octave* is sometimes referred to when discussing signals and measurements. An *octave* means that the fundamental frequency of one signal is twice (double) the fundamental frequency of another signal. For example, *one octave above* a tone with 200 Hz fundamental frequency is 400 Hz. Likewise, a tone with frequency 800 Hz is said to be *two octaves higher* than a 200 Hz tone: one octave above 200 Hz is 400 Hz, and another octave (frequency doubling) gives 800 Hz.

2.4 Wave Propagation and Spherical Spreading

Sound waves in air propagate away from the source in all directions. Sources with small dimensions compared to the sound wavelength cause approximately equal sound pressure waves emanating in all directions from the source, which is referred to as *spherical* wave propagation. As the sound propagates spherically outward, the sound energy from the source is distributed over the ever-increasing surface of the

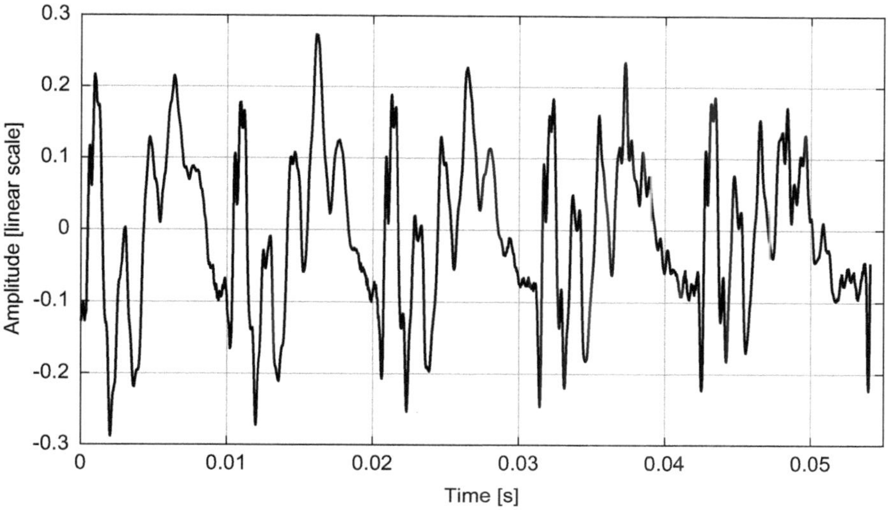

Fig. 2.8 A quasiperiodic section of approximately 5 cycles of a recorded speech signal

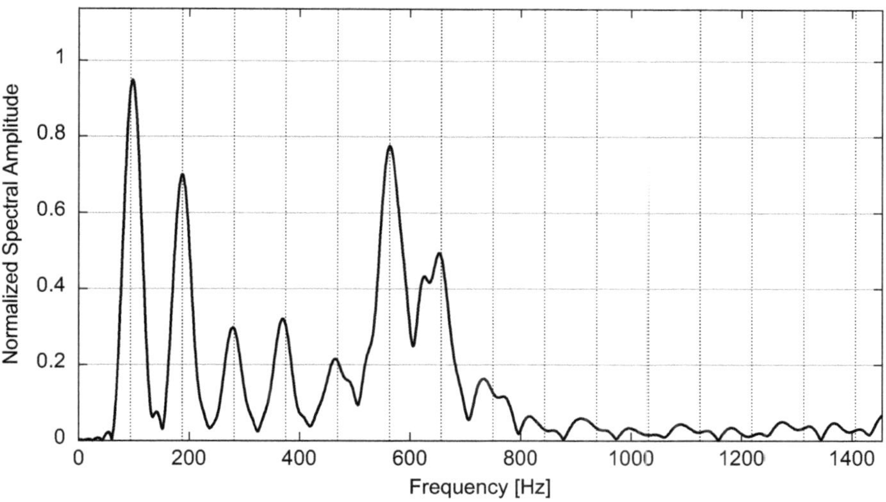

Fig. 2.9 Fourier transform magnitude of the quasiperiodic speech waveform of Fig. 2.8. Vertical grid depicts approximate harmonic spacing of 93.75 Hz

growing sphere, meaning that the sound power in a particular direction decreases with increasing radius. Theoretically, if no sound reflections are present, the spherical surface area is proportional to the radius (r) squared (surface area = $4\pi r^2$), so the wave *intensity* (watts per unit area) decreases as $1/(r^2)$. The acoustic intensity, in turn, is proportional to the square of the acoustic pressure, so the wave pressure amplitude decreases as $1/r$, again neglecting the presence of any sound reflections.

The practical result is the well-known observation that a sound source becomes quieter as the observation distance increases (Kinsler et al. 2000).

The 1/radius factor for the decrease in acoustical pressure with increasing distance results in a reduction of 6 decibels SPL for each doubling of distance: $20 \log_{10}((1/r) \times P/P_{ref}) = 20 \log_{10}(P/P_{ref}) - 20 \log_{10}(r)$, and if r changes from 1 to 2 (distance doubles), $-20 \log_{10}(2) = -6.02$ dB. However, in most practical circumstances, the spherical wave propagation encounters boundary surfaces, such as the ground, walls, and other physical obstacles, so the acoustic pressure will deviate from the simple spherical spreading prediction due to the superposition of the direct sound path and reflected waves from the surfaces (see Sect. 2.4.1).

It turns out that the speed of sound in air varies with the temperature of the air: sound travels faster in warm air and slower in cold air. Thus, for a given frequency, the wavelength will be *longer* in warm air where the speed is faster and *shorter* in cold air where the speed is slower (Fig. 2.10).

In most circumstances, the temperature dependence of the speed of sound is not noticed in any practical way, except outdoors when different layers of air have differing temperatures (Kinsler et al. 2000). Differing air layer temperatures mean different sound speeds. For example, on a cold winter morning, the air will often exhibit a temperature *gradient*, with colder air near the ground and warmer air aloft. In this situation, the sound waves travel slower when close to the ground and faster in the warmer air above. Thus, a sound wave front moving horizontally over the ground will exhibit *downward refraction*: the portion of the sound wave front that moves through the cold surface air travels more slowly than the portion of the wave front that moves through the warmer air above, causing the sound wave front to bend *downward*.

Conversely, in the early evening after a warm sunny day, the air temperature near the sun-warmed surface of the ground will be higher than the air temperature aloft.

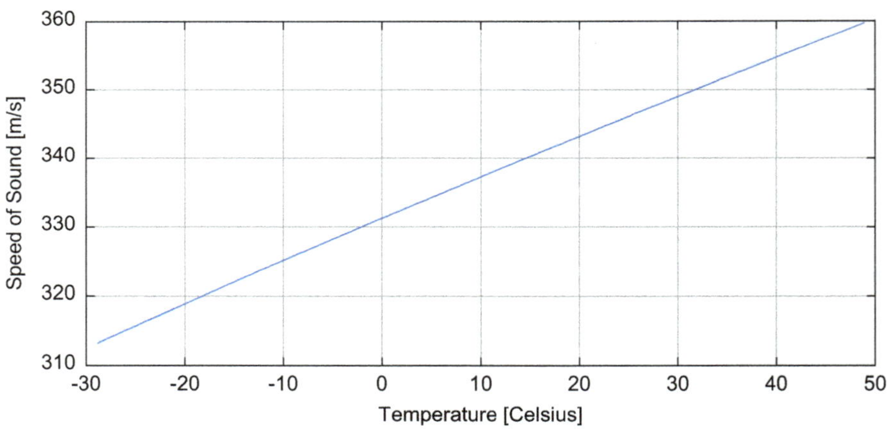

Fig. 2.10 Speed of sound in air as a function of air temperature

This means that the sound wave front will be *refracted upward* due to the wave front moving faster in the warm air layer near the ground compared to the portion of the sound moving in the cooler, slower air higher up.

The result (see Fig. 2.11) is that a distant sound may be heard more easily on a cold morning due to the refractive focusing effect, while a distant sound may be less audible when the air near the surface is hot, due to the upward bending of the wave front.

Along with spherical spreading and the potential refraction effects, sound propagation may have energy losses due to humidity and temperature of the air (Harris 1966). These aspects of sound physics can have implications in audio forensic analysis, particularly for outdoor sounds observed at a significant distance from the source.

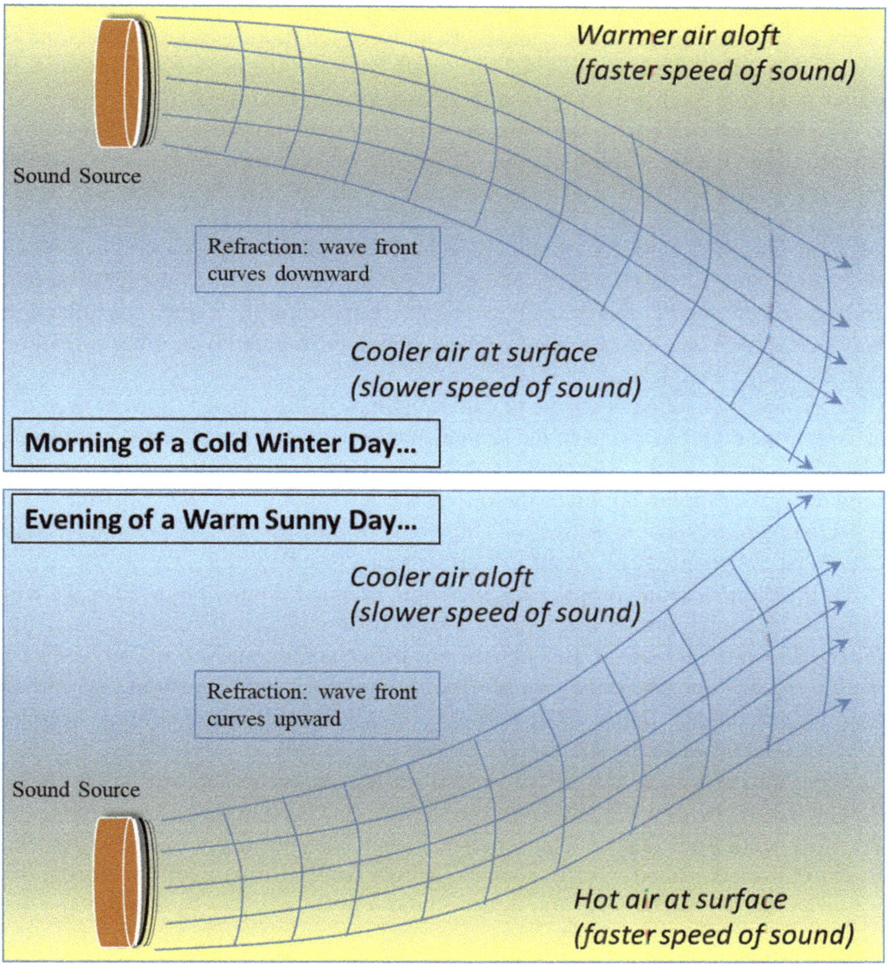

Fig. 2.11 Refraction—cold air near the ground and warmer air aloft causes the sound wave front to curve downward. Warm air near the ground and colder air aloft causes the wave front to curve upward

2.4.1 Reflections and Reverberation

A microphone detects the instantaneous acoustic pressure, which consists of the sound waves propagating directly through the air from a sound source to the microphone position, sound waves that reach the microphone after reflecting from the ground, walls, and other surfaces, and *reverberant* sound that arrives after multiple surface reflections. Therefore, an acoustic recording will contain information about the sound source *and* the acoustical properties of the physical surroundings in which the recording is made. If there are distinctive *background* sounds, such as mechanical noises, music, or alarm chimes, these sounds will also be detected by the microphone along with the more prominent *foreground* sounds.

The relative time of arrival of the direct sound and a reflection of that sound depends upon the difference in path length between the sound source and the microphone. Assuming there is a line-of-sight path between the source and the microphone, the direct path will always be the shortest path (see Fig. 2.12).

The most noticeable reflections occur when the reflecting surface is relatively far from the source and the microphone. The greater reflection distance means that the reflected sound will arrive at the microphone substantially later than the direct sound, resulting in an audible *echo*. On the other hand, if the source and microphone are both relatively near the reflecting surface, such as the source and microphone both being close to the ground in an open area, the delay between the direct sound and the reflection will be very small and may be imperceptible. Even if the reflection is imperceptible to a listener, it may be detectable in an audio recording, as will be discussed later in this book.

If the speed of sound is known or can be estimated, the elapsed time between the arrival of the direct sound from the source and the arrival of the reflected sound acts like a "measuring stick." The product of the speed of sound [meters/second], and the time difference [seconds], gives the difference in length of the direct path and the reflected path. In some investigations, this type of information can potentially help determine the geometry of events at a crime scene (see Sect. 7.2.1).

As mentioned, sound propagating in a room or in some other bounded space will result in the microphone receiving a superposition of the direct sound and the reflected/reverberant sound. In a large room with a continuous sound source, like a lecture speaker or a musical ensemble, the *reverberant* sound energy in the room is found to be roughly uniformly distributed. The reverberant sound reflections come from all directions, so the sound field is described as *diffuse*.

If the microphone is close to the sound source, the recording will ordinarily be dominated by the direct sound from the source compared to the reverberation. On the other hand, if the microphone is moved to a greater distance from the source, the level of the background reverberation in the room will be roughly the same, but the sound pressure amplitude of the direct sound will be reduced because of the $1/r$

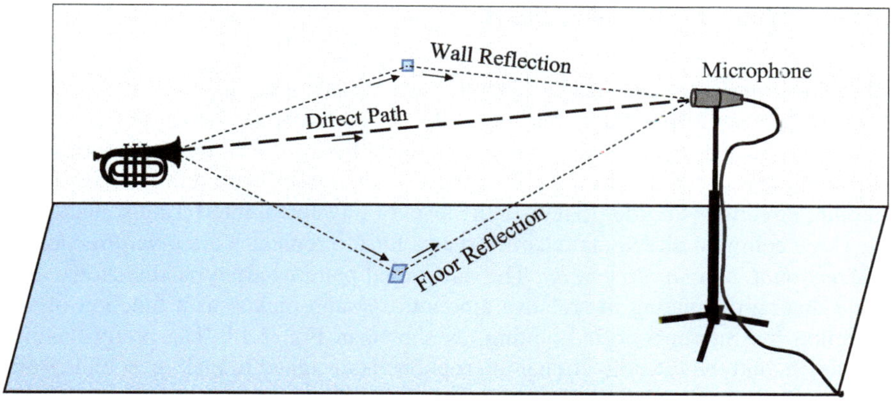

Fig. 2.12 Direct sound and first-order reflected sound received by a microphone

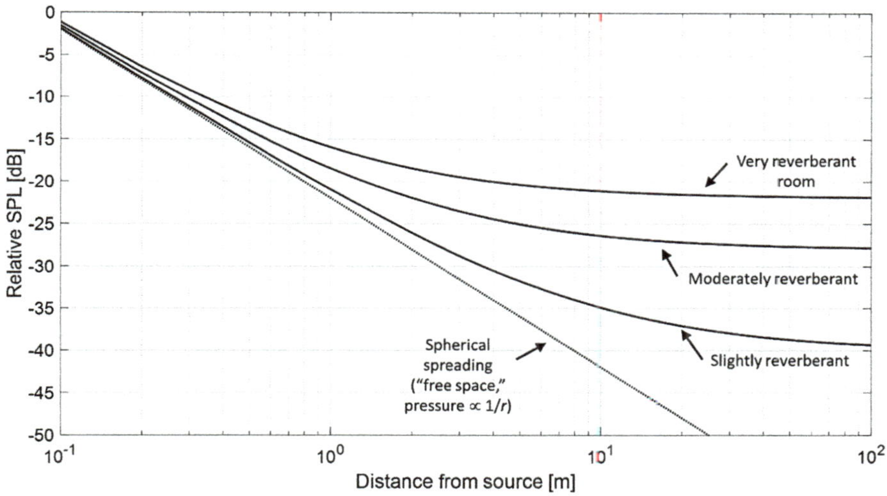

Fig. 2.13 Example of sound pressure level as the microphone is moved farther from a continuous source in a reverberant room. If no reverberation is present, the relative sound level follows the inverse 1/r effect (−6 dB/doubling distance) due to spherical spreading, but if the room is reverberant, the sound level reaches the background reverberation level as the microphone moves farther from the source

effect of spherical propagation. The effect is that the balance between the direct sound and the room reverberation in the recording will change from being dominated by the direct sound to being dominated by the reverberation as the distance between the source and the microphone increases. This effect is shown in Fig. 2.13.

2.4.2 Microphone Directionality

The *directional characteristics* of the microphone also play a role in the recorded signal. Microphones generally respond to the acoustic pressure upon the diaphragm, but microphone designers may choose to engineer the device deliberately to respond preferentially to waves arriving from particular directions or to minimize response in other directions in order to reduce the level of unwanted, interfering sounds.

Three common directional characteristics for microphones are *omnidirectional*, *bidirectional*, and *unidirectional*. The directional patterns are typically shown as a polar diagram depicting the relative amount of sound pickup as a function of the direction the microphone is pointing, as shown in Fig. 2.14. The prefix "omni" means *all*, and the *omni*directional microphone is designed to pick up sounds from all directions. Thus, its directional pattern is equal in all directions, making a circle. The *bi*directional microphone picks up sound principally from two directions: the direction the microphone is pointing (0°) and from the opposite direction (180°), but it does not pick up sound off to the side of the microphone. Its directional pattern is therefore unity in the forward (0°) and backward (180°) directions and zero in the sideways direction (90° and 270°). The *uni*directional microphone picks up sound in one direction, the direction the microphone is pointing (0°). The unidirectional mic is designed to be insensitive to sounds coming from the opposite direction (180°).

The bidirectional microphone is sometimes called a "figure eight" microphone, because its directional pattern looks like the number 8. The unidirectional microphone is often called a *cardioid* mic, because it's somewhat heart-shaped directional pattern traces a mathematical cardioid.

If a directional microphone is pointed (0°) toward a sound source, the recorded signal level will be relatively high, compared to the signal if the sound source is off to the side of the microphone at an angle where the directional mic is less sensitive. Thus, two recorded sounds that have different levels in a forensic recording might be from different sources or could be from the same source if there was movement of the microphone or the source so that the direction was changed.

Similarly, if a directional microphone is pointed toward the sound source in a reverberant room, the balance in the recording will emphasize the direct sound from the source compared to the reverberant sound in the room. This is because the reverberant sound arriving from off-axis directions is attenuated by of the directional mic's spatial selectivity compared to the direct sound from the source.

2.5 Human Hearing Characteristics

Audio forensic investigations may involve questions of *audibility*: would a particular sound be audible by a listener under the circumstances presented? For example, a question may arise about whether an alarm signal was audible at a certain

Fig. 2.14 Directional pattern diagrams for three common microphone types: (**a**) omnidirectional, (**b**) bidirectional, and (**c**) unidirectional. The diagram indicates the relative sound pickup as a function of angle with respect to the direction the microphone is pointing

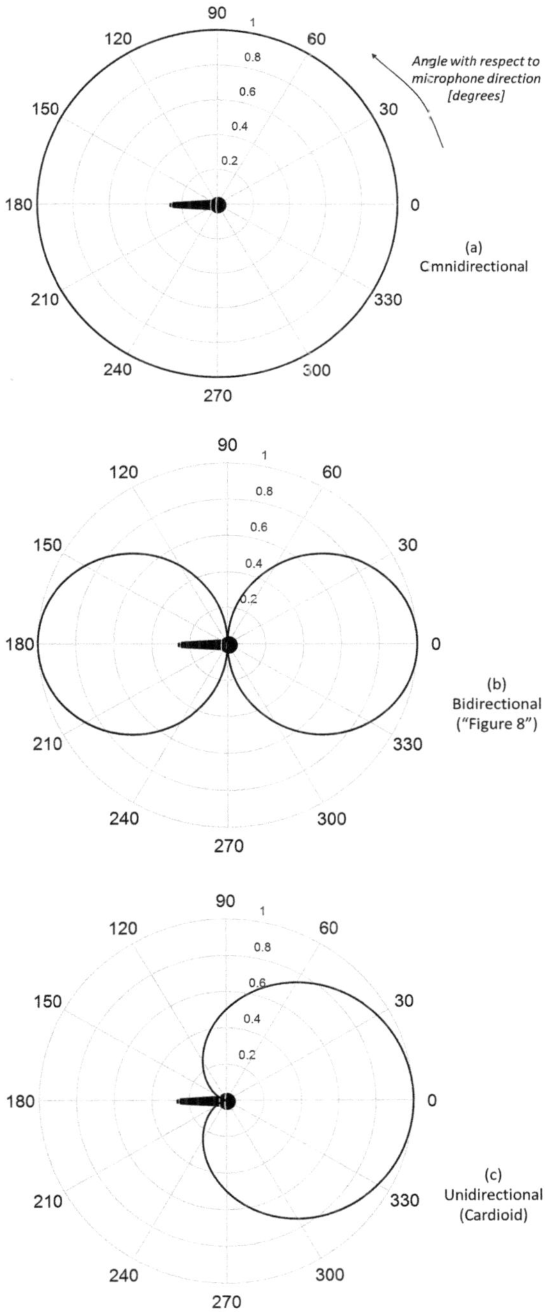

distance, or in the presence of known interfering noise. These sorts of questions require an understanding of the strengths and weaknesses of the human hearing system. The next two subsections provide a brief, simplified view of (1) the anatomy and physiology of the auditory system comprised of the ear and its neural connections to the brain and (2) subjective aspects of hearing (psychoacoustics) and the ability to recognize a signal of interest in the presence of competing sounds and noise.

2.5.1 Anatomy and Physiology of the Ear

The ear is the sensory organ related to hearing that transduces sound energy into a neural code that is processed by specialized structures of the brain. The basic anatomy of the ear comprises three sections: the outer ear, the middle ear, and the inner ear (Kinsler et al. 2000; Pickles 2013).

The outer ear is the externally visible portion of the auditory system. The external ear flap, known as the *pinna*, surrounds the opening of the external auditory canal. Many mammals, such as a deer or a cat, have moveable pinnae that the animal can rotate deliberately in a particular direction. Human pinnae, however, are not generally moveable in any practical way, except by rotating the entire head. The *concha* refers to the central depression in the pinna that is connected to the external opening of the *auditory canal*, or *ear canal*.

The ear canal, slightly curved along its midline, is approximately 0.8 cm in diameter and 2.5 cm in length. It is exposed to the air outside the head as it connects to the concha, so the average air pressure in the canal is equal to the ambient air pressure outside the head. The inner end of the auditory canal is completely sealed by the airtight and watertight *tympanic membrane* (the *eardrum*). The ear canal helps protect the eardrum and the other sensitive structures of the middle and inner ear while still allowing direct acoustical coupling of external sound.

The shape and structure of the external ear, and the position of the ears with respect to the head and upper body, cause acoustic *diffraction* that depends upon the direction (azimuth and elevation) of the sound source and the wavelength of the sound. Most of the ability to *localize* a sound source in the azimuthal (left-right) plane depends upon *binaural* hearing: the sensory apparatus of each ear encodes sounds independently prior to the higher levels of combined neural processing.

When listening to sounds in free-field or with circumaural (enclose the ear) or supra-aural (on-the-ear) headphones, the acoustical pathway from the concha to the eardrum is employed, but when listening with insert headphones (earbuds), the concha part of the external ear acoustical path is not used.

The *middle ear* is located between the eardrum and the inner ear and is comprised of three small *ossicles* (tiny bones) and a middle ear *tympanic cavity* that is located deep within the temporal bone of the human skull. The ossicles are the *malleus* (or hammer) connected to the inner surface of the eardrum, the *incus* (or anvil), and the *stapes* (or stirrup), which is connected to the *oval window* of the inner ear

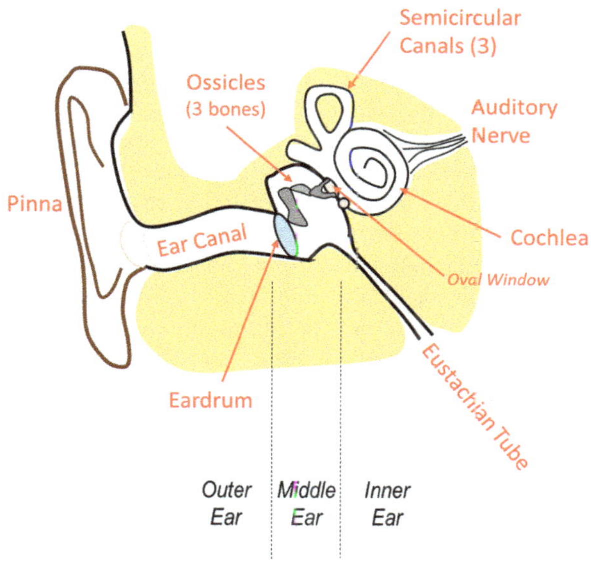

Fig. 2.15 Simplified view of the anatomy of the human ear

(see Fig. 2.15). The stapes is the smallest and lightest bone in the human body. The ossicles are suspended together within the middle ear cavity via ligaments and two tiny muscles, the *tensor tympani* and the *stapedius*. The tensor tympani is connected to the malleus (hammer), while the stapedius, the smallest skeletal muscle in the human body at approximately 1 mm in length, is attached to the stapes. These auditory muscles play a role in a physiological reaction known as the *acoustic reflex*, which will be described in the next subsection.

The *Eustachian tube* (see Fig. 2.15) is a canal between the middle ear cavity and the back of the nasopharynx, the junction region between the throat and the nasal cavity. Normally the Eustachian tube (of each ear) is closed, although it briefly opens during swallowing to allow gradual movement of air in or out of the middle ear through the Eustachian tube. If there is a steady air pressure difference between the air trapped in the middle ear and the air in the nasopharynx, the tube allows air to seep through and equalize the pressure on the exterior and interior sides of the eardrum. If there is a sudden change in ambient pressure, like when ascending or descending quickly in an airplane, the more rapid air movement through the Eustachian tubes gives rise to the familiar "ear popping" sensation.

The air in the middle ear cavity is ordinarily at approximately the same ambient pressure as the air outside the head. Sometimes, the air pressure in the middle ear cavity may differ from the ambient atmospheric pressure due to temporary blockage of the Eustachian tube, inflammation, or illness. This creates a pressure imbalance between the exterior and interior sides of the eardrum, which stiffens the membrane and alters the mechanical sensitivity of the ossicle chain.

The *inner ear* is composed of the *cochlea* as the organ of hearing, and three *semicircular canals* and related structures that form the vestibular organ of *balance*.

While not of significance to audio forensic analysis, the three semicircular canals are sensitive to angular accelerations in each of three dimensions in space, and two smaller vestibular structures are sensitive to linear accelerations in relation to the action of gravity. Neural encoding of such motions forms the physiological basis underlying a person's sense of balance and spatial orientation, and the ability to integrate physical movement with balance.

The cochlea is the primary neurosensory organ of the hearing system. The cochlea is a bony cavity of spiral shape that encases and protects the soft biological tissues within that are exquisitely sensitive to sound-induced vibrations. The interior of the cochlea is divided into several fluid-filled chambers and microscopic neural structures. The stapes bone of the middle ear attaches to the oval window of the cochlea. The detection of sound-induced vibrations and the conversion into the neural code occurs within microscopic *hair cells* arranged on the *organ of Corti* inside the cochlear structure. There are approximately 3500 *inner* hair cells and about 12,000 *outer* hair cells within the human cochlea. The inner hair cells provide neural transduction of the vibratory stimuli, while the outer hair cells are thought to function as a cochlear amplifier and gain compressor. The neurons of the auditory nerve connect from the base of each inner hair cell to locations in the brainstem.

The stages of processing in the auditory physiological pathway are briefly sketched in the remainder of this section. Because sound is a small pressure fluctuation above and below the ambient pressure, the presence of sound waves in the air around the head means that the instantaneous pressure in the ear canals will be alternately higher and lower than the fixed air pressure held in the middle ear cavity. The acoustic pressure difference between the air in the ear canal and middle ear causes a net force on the eardrum, making it move alternately in and out in response to the pressure difference. The greater the sound pressure amplitude, the greater the force on the eardrum and the greater the in-and-out displacement. Oscillatory sound energy that displaces the eardrum generates motion of the malleus and other ossicles. Thus, the middle ear acts as a *transducer* to convert acoustic energy into mechanical energy via oscillatory transmission of forces and torques through the ossicular chain. To a first approximation, the ossicular chain acts as a mechanical lever system, conveying the force on the relatively large eardrum down to the tiny oval window aperture and into the mechanical structures of the fluid-filled cochlea.

An important function of the middle ear is to transmit energy efficiently from the sound waves in air to a mechanical displacement of the sensory structures within the cochlea. The middle ear acts as an impedance transformer in terms of the mechanical level action, but more importantly, the sound energy arriving at the relatively large area of the pinna is delivered to the intermediate area of the eardrum and thence concentrated to the very small area of the stapes footplate. In summary, the external and middle ear structures mitigate to a large degree the acoustic impedance mismatch between the pressure and particle velocity in air and the corresponding mechanical displacements in the cochlea (Allen et al. 2005; Pickles 2013).

The foregoing is only a very brief tour of the auditory pathway for a single ear. This pathway exists for each of our two ears, and specialized brain regions exist at which neural information is combined across both ears for binaural processing. The higher-level processing in the auditory system includes nerves and structures that share information between the two ears, enabling our ability to estimate the direction and distance of a particular sound source relative to one's head. For interested readers, much more detailed explanations of auditory anatomy and physiology are available (Geisler 1998; Pickles 2013).

2.5.2 *Psychoacoustics*

The human auditory system has many significant strengths and weaknesses as a detector of sound. The sensation of hearing is commonly understood to have a frequency range from approximately 20 Hz to approximately 20 kHz when measured under laboratory conditions, but the ear's ability to detect sound depends upon the pressure amplitude at a given frequency, the complexity of the stimulus, and factors that vary from one listener to another. Although beyond the scope of this book, there are many interesting and accessible references regarding the human auditory system, and interested readers are invited to study this fascinating field (Bess and Humes 2008; Moore 2012).

Audio forensic examiners do not commonly address the physiology of the human hearing system, but sometimes the *perception* of sound becomes important for cases involving questions of audibility, intelligibility, speaker identification, and other earwitness testimony (Koenig 1986). There are many interesting aspects to human psychoacoustics, but for the purposes of this book, we will focus upon only three: frequency sensitivity, frequency masking, and speech detection in noise.

Unlike *sound pressure level*, which has a precise objective definition, sound *loudness* is a perceptual quantity that depends upon the listener. Large tests of human subjects demonstrate that our subjective judgment of sound loudness depends upon both the frequency and the amplitude of the sound at our ears. Acousticians use empirical charts of *equal-loudness contours*, such as the Fletcher-Munson or Robinson-Dadson graphs, to show the average sensitivity behavior. In these studies, the researchers recruited a large number of young, healthy people to perform a subjective loudness test. The subjects heard a sinewave tone at 1 kHz at a fixed sound pressure level and then turned a knob to adjust the loudness of a tone at some other frequency until they felt the tone was equally loud as the 1 kHz reference tone. This process was repeated for a range of sound pressure levels for the 1 kHz reference, and the researchers averaged the performance over all of the people participating in the test.

The resulting average response (see Fig. 2.16) shows that young, healthy listeners generally need a higher sound pressure at frequencies below 1 kHz in order for the signal to be judged equally loud as the 1 kHz tone: the typical healthy ear is *not as sensitive to low-frequency sounds* compared to tones in the 2–4 kHz range (ISO 2003). The best sensitivity is found for sound frequencies around 3 kHz, which corresponds to the wavelength that excites the tube resonance of the auditory canal.

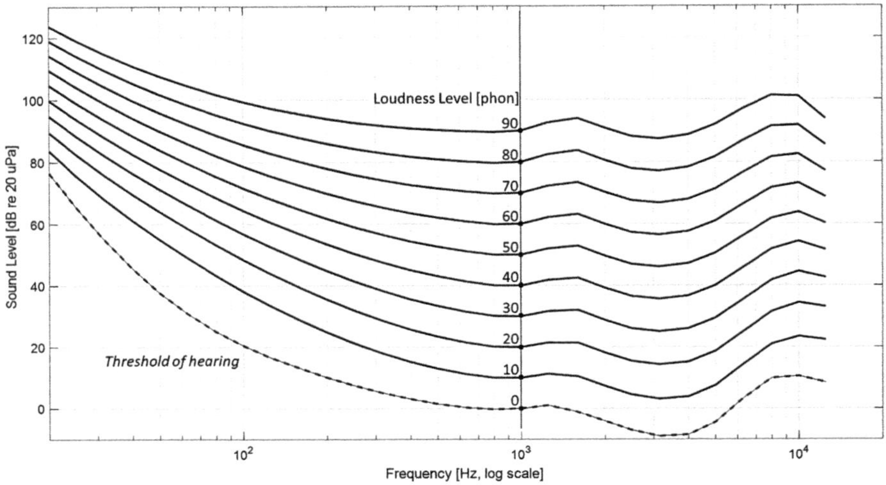

Fig. 2.16 Equal-loudness contours for human hearing based on International Organization for Standardization standard 226:2003 (ISO 2003)

The average ear is also somewhat less sensitive at frequencies above 4 kHz, until having little or no sensation as the frequency exceeds 20 kHz.

Another very important observation about human hearing we can see in the equal-loudness curves is that the sensitivity not only varies with frequency but *also varies with amplitude*: the equal-loudness curves are *flatter* (more consistent sensitivity) as the 1 kHz reference loudness increases. In other words, we judge louder tones at all frequencies to be more equal in perceived loudness than if we compare quiet tones at various frequencies.

Besides the sensitivity vs. frequency effects, the human hearing system also exhibits temporal changes in sensitivity when exposed to loud sounds. The *acoustic reflex* is a physiological neural response of the ear induced by a high-level sound (e.g., a gunshot). The tiny stapedius muscle contracts, altering the mechanical coupling of the stapes footplate to the oval window of the cochlea. This contraction protects the inner ear somewhat from the possibly damaging effects of loud sounds. Like other muscular somatic reflexes in the body, the acoustic reflex is not consciously controlled, but the effect can be a change in level sensitivity of perhaps as much as 15–20 dB. However, the acoustic reflex takes time to react to loud sounds, so we cannot count on any significant protection from abrupt and impulsive sounds such as nearby gunshots. The stapedius muscle gradually returns to its normal state when the loud sound exposure ceases.

As noted earlier, the equal-loudness characteristics of Fig. 2.16 were obtained from a population of young, healthy listeners. Individual listeners may have notable differences from these nominal curves, especially if the middle and inner ear structures have been subjected to injury from noise exposure, disease, or neurological damage. Some individuals have significant differences in sensitivity between the two ears. Perhaps the most important observation is that hearing sensitivity almost always decreases with increasing age. Age-related hearing sensitivity loss, known as *presbycusis*, typically occurs gradually, and so an individual may not notice the effects immediately when the natural changes start to occur. A physician specializing in understanding disorders of the ear, nose, and throat (ENT), known as an *otolaryngologist*, can be consulted for advice on hearing issues, and an *audiologist* can provide periodic tests to measure hearing sensitivity.

The audio forensics lesson here is that the ear is a *nonlinear* and *time-variant* detector, and we need to be careful to interpret forensic results for both earwitness testimony AND for the examiner using his or her ears to interpret audio evidence in the lab (Maher 2015). It is highly recommended that forensic examiners have periodic hearing screening tests to track any changes in hearing acuity. As will be described in Chap. 4, an audio forensic examination involves more than just listening, but the examiner's hearing sense inevitably plays a key role in most forensic investigations.

Masking is the term used to describe the phenomenon that the ear and brain may have more difficulty noticing the presence of a particular sound when there are other sounds presented simultaneously, or nearly simultaneously, with nearly coincident frequency content (Moore 2012). In fact, a sound that is clearly detectable when presented on its own may become perceptually inaudible—*masked*—in the presence of other sounds with particular frequency content and sound level. At the extremes, it is quite familiar to have a loud television drown out the sound of light knock on the door or to have a party full of loud simultaneous conversations and background music interfere with your own conversation. However, the masking effect can also occur with relatively quiet sounds and circumstances.

While the masking effect may be a sign of annoyance if one cannot hear a desired conversation in the presence of noise, the effect is also helpful for estimating the ability of the human hearing system to detect unwanted or irrelevant background sounds. For example, *frequency masking* is exploited in most contemporary perceptual audio coding systems (e.g., MP3, AAC, WMA) by allowing the level of coding noise (signal discrepancy) to increase in frequency bands in which the noise will be masked effectively by stronger components of the recorded signal itself. The discrepancy is present in the signal, but if the algorithm designer did a good job, the signal defect is *inaudible* to a *human listener*. This means that the carefully encoded audio can use fewer bits to represent an acceptable replica of the original audio signal for human listening. However, a forensic audio examiner must be careful when attempting to interpret the reconstructed signal's waveform and spectrum using objective measurements and calculations: the perceptual encoding may have introduced signal features that, while possibly inaudible to a listener, may change or interfere with objective analysis. See Sect. 2.8 for more information.

2.5.3 *Frequency Weighting in SPL Measurements*

Because the human ear has nonuniform sensitivity as a function of frequency, sound pressure level measurements often use a filter that approximates the ear's sensitivity. The filter is known as a *weighting filter* because it emphasizes (weights) the sound

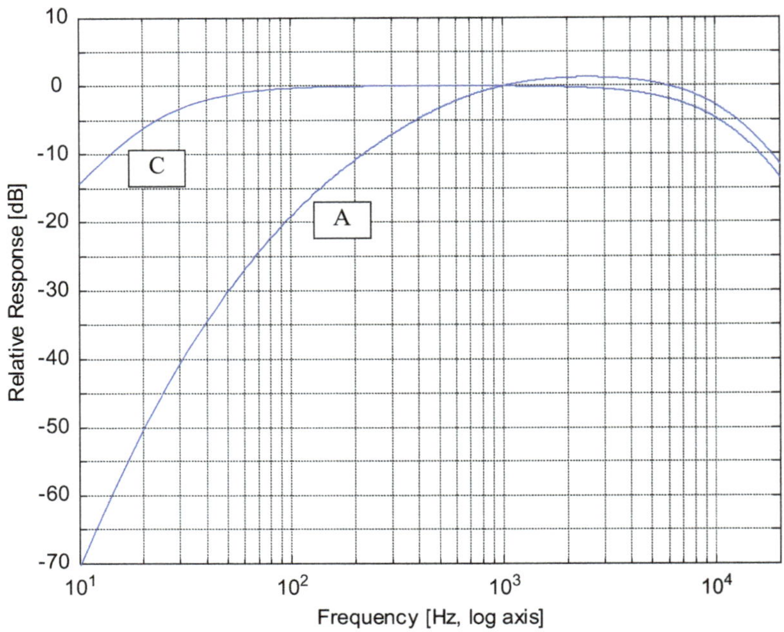

Fig. 2.17 Customary "weighting" filters A and C used for sound level measurements

energy in the frequency range for which the ear is most sensitive, while weighting less the ranges where the ear is less sensitive. The resulting filter is a *bandpass* filter, meaning that it primarily passes the portion of the signal within a particular frequency range, or *band*. The most common weighting filter is the standardized *A-weighting* filter, which approximates the average equal-loudness curve for a 40 dB reference signal. Standard sound level meters usually have an A-weighting setting, and some may have other weighting options, such as the C-weighting and an "unweighted" (flat) frequency selection. If a weighting filter is used to make a sound level measurement, the reading should be specified, e.g., "the meter reading was 45 dBA re 20 μPa," where the "dBA" indicates that the A-weighting filter was in use (Kinsler et al. 2000) (Fig. 2.17).

2.5.4 Speech Intelligibility

Audio forensic examinations often involve interpreting audio recordings containing human speech. In some cases, the request may be to assess the likelihood that a speech utterance was intelligible under the conditions described by the witness or established by other evidence.

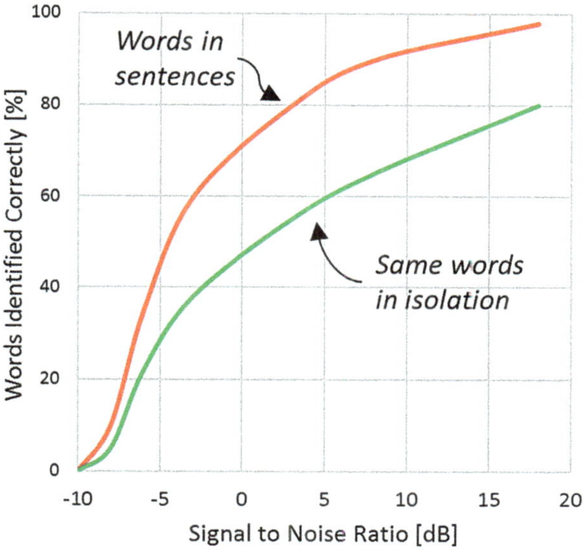

Fig. 2.18 Intelligibility of speech for sentences and isolated words (after Miller et al. 1951)

As a primary means of communication, human speech has evolved to contain significant *redundancy* so that a listener is likely to understand the talker's remarks even in the presence of competing sounds and noise. The structure of languages also provides context and semantics that allow the listener to get the gist of a statement without necessarily understanding every word. Nevertheless, noise tends to interfere with the intelligibility of speech communication (Quatieri 2002).

Noisy speech is often described by a *signal-to-noise ratio* (SNR) expressed in decibels. The SNR is usually estimated using assumptions about the speech level and the level of the interfering noise. A signal with 0 dB SNR means that the signal (speech) level and the noise level are the same, while a negative dB SNR means that the noise is at a higher level than the speech.

Subjective tests of the intelligibility of noisy speech usually follow the behavior shown in Fig. 2.18. The intelligibility (percent correct for a listener transcribing a conversation) is virtually 100% for SNRs above 10 dB and quickly falls essentially to zero percent intelligible when the SNR is worse than −10 dB.

Human speech has significant signal energy (bandwidth) of roughly 200 Hz–4 kHz. This is the audio bandwidth transmitted by common telephones and mobile radio systems intended for speech messages. Increasing the audio bandwidth generally results in listeners happy with the improved *quality* of the speech, but the *intelligibility* does not necessarily improve even if listeners perceive that the quality is to be better. This fact is important to remember when an audio forensics expert is asked to improve the quality of a noisy speech recording: sometimes a processed recording can have lower speech intelligibility even if listeners think it sounds better in quality. Several examples of this phenomenon are discussed in Sect. 6.2.

2.6 Signal Processing

Like the human ear, audio engineering systems take the sound pressure fluctuations in the air and convert the acoustic energy into mechanical motion and electrical signals. Physicists and engineers refer to the conversion of energy from one form to another as *transduction*, and audio *transducers* include microphones and loudspeakers.

Microphones include a diaphragm similar in function to the eardrum: the instantaneous air pressure on the side of the diaphragm exposed to the sound source differs from the fixed air pressure on the other side of the diaphragm, causing a differential force to move the diaphragm in and out with each sound pressure cycle. The mechanical motion of the diaphragm drives a generating element that converts the motion into an electrical signal. Over the years, audio engineers devised many different generating elements for use in microphones: variable resistance, electromagnetic induction, variable capacitance, piezoelectric material, etc.

The electrical signal generated by the microphone in response to sound is an *analog* signal: the continuous variation in the electricity with time is linearly proportional to the continuous variation in the acoustic pressure wave impinging upon the diaphragm, so the electrical signal is *analogous* to the pressure signal. The analog audio signal can be amplified, filtered, recorded, reproduced, modulated, broadcasted, and otherwise processed like any other electrical communications signal.

Loudspeakers perform the complementary transduction of analog electrical signals into sound. The usual design of a loudspeaker driver includes a *motor* element that produces a force and motion proportional to the audio electrical signal and a *diaphragm* that efficiently transfers the mechanical motion of the motor into acoustic waves. Loudspeakers generally comprise a system in which the driver (motor and diaphragm) operates within a specially constructed resonant enclosure (cabinet) that helps improve the linearity and efficiency of the speaker system. The enclosure, the driver, and even the amplifier in the case of powered speakers are designed together. Modern loudspeakers often utilize multiple drivers of different sizes in order to optimize the reproduced sound over the extremely wide range of wavelengths accommodated by audible sounds (wavelength greater than 17 m for 20 Hz, down to less than 2 cm for 20 kHz).

2.7 Digital Audio

While a microphone produces an analog signal, contemporary audio systems almost exclusively involve *digital* signal processing and storage. *Digitization* refers to two processes performed by a circuit known as an *analog-to-digital converter* (ADC). The first process in the ADC is *time sampling*, which means a rapid and repeated measurement of the instantaneous value of the analog audio signal many times per second. Each individual measurement is a time *sample*. The rate at which the time sampling occurs is called the *sampling rate*, expressed in samples per second [Hz]. The second process in the ADC is *quantization*, which means representing each waveform sample with an integer value. The *precision* of the measurement is typically expressed by the number of digital bits used for each sample. For example, using the most recognized representation known as pulse-code modulation (PCM), a 16-bit quantization represents each sample's amplitude using a 16-bit integer, which can be one of $2^{16} = 65{,}536$ different values (−32,768 to 32,767).

For example, the standard audio compact disc (CD) has two audio channels (stereo), each sampled at 44.1 kHz sampling rate, and 16-bit resolution for each sample.

Unlike analog signals, digital signals can be stored in computer memory, transmitted over digital networks, and protected with error-correcting coding. What's more, perfect copies can be made of a digital recording. However, care must be taken to ensure that the digitization allows sufficient audio bandwidth by using a sufficiently fast sampling rate and also allows sufficient amplitude precision by using a sufficient number of bits in the quantizer. The mathematical theory of digital sampling requires that the sampling rate be at least twice the bandwidth of the analog signal being sampled (the *Nyquist* rate), meaning that the sampling rate will exceed 40 kHz to accommodate the entire 20 kHz audible bandwidth. The quantization precision is usually determined by the required signal-to-quantization noise ratio (SQNR) required for a particular application. Telephone-quality speech may use 8-bit or 12-bit quantization (45–75 dB SQNR), while high fidelity music will generally need at least 16-bit quantization (>90 dB SQNR).

The complementary process, the *digital-to-analog converter* (DAC), reconstructs the analog signal from its digital representation. Reconstruction is typically the last step before the power amplifier that drives the loudspeaker or headphones used for listening.

2.8 Perceptual Audio Coding

The traditional audio sampling, quantization, and reconstruction process described above works well but results in a *bitrate* [bits/sec = (bits/sample) × (samples/sec)] that is too high for small and inexpensive transmission and storage systems. Since the late 1980s, digital audio signal processing systems exploiting the strengths and weaknesses of the human hearing system provide very good *perceptual* quality at bitrates much lower than the (bits/sample) × (samples/sec) of a traditional digital audio system. Perceptual audio coding algorithms, such as MP3 [MPEG (Moving Picture Experts Group) 1, Layer 3], Dolby Digital, and MPEG Advanced Audio Coding (AAC), rely upon the *masking* phenomenon of human psychoacoustics to use a lower bit rate while still concealing the high level of quantization noise during time intervals with strong signal components. While the reconstructed audio has good quality for human listeners, it is important to understand that the perceptual audio coding systems are *lossy* coders. This means that the unlike a traditional digital audio system in which the discrepancy between the original signal and the reconstructed signal is bounded by the quantization level, the waveform discrepancy for the perceptually encoded signal can be much greater in magnitude, even if the differences are inaudible to a human listener.

Audio forensic examination increasingly involves recording systems that produce perceptually encoded audio, and care must be taken when applying waveform analysis when lossy coding is in use. A related area of concern occurs when decoding a lossy-coded signal and then compressing it again by another lossy re-encoding. Even if the second encoding uses the same algorithm as the original encode/decode, the sequence of lossy compression, reconstruction, lossy compression, reconstruction, etc. will cause the accumulation of audible artifacts and distortion. Generally, *perceptually encoded audio should never be equalized or re-encoded*, as these processes change the spectral details exploited by the perceptual encoding algorithms.

References

Allen, J. B., Jeng, P. S., & Levitt, H. (2005). Evaluation of human middle ear function via an acoustic power assessment. *Journal of Rehabilitation Research and Development, 42*(4), 63–78.

Allen, J. B., & Rabiner, L. R. (1977). A unified approach to short time Fourier analysis and synthesis. *Proceedings of the IEEE, 65*(11), 1558–1564.

Bess, F. H., & Humes, L. E. (2008). *Audiology: The fundamentals* (4th ed.). Philadelphia, PA: Lippincott Williams and Wilkins.

Geisler, C. D. (1998). *From sound to synapse: Physiology of the mammalian ear*. New York: Oxford University Press.

Harris, C. (1966). Absorption of sound in the air versus humidity and temperature. *The Journal of the Acoustical Society of America, 40*(1), 148–159.

Hartmann, W. M. (2013). *Principles of musical acoustics*. New York: Springer.

International Organization for Standardization (ISO). (2003). *ISO 226:2003: Acoustics-normal equal-loudness-level contours*. Geneva, Switzerland.

Kinsler, L. E., Frey, A. R., Coppens, A. B., & Sanders, J. V. (2000). *Fundamentals of acoustics* (4th ed.). Hoboken, NJ: Wiley.

Koenig, B. E. (1986). Spectrographic voice identification: A forensic survey. *The Journal of the Acoustical Society of America, 79*, 2088–2091.

Maher, R. C. (2015). Lending an ear in the courtroom: Forensic acoustics. *Acoustics Today, **11***(3), 22–29.

Miller, G. A., Heise, G. A., & Lichten, W. (1951). The intelligibility of speech as a function of the context of the test materials, *Journal of Experimental Psychology, 41*(5), 329–335.

Moore, B. C. J. (2012). *An introduction to the psychology of hearing* (6th ed.). Bingley, UK: Emerald Group Publishing.

Pickles, J. O. (2013). *An introduction to the physiology of hearing* (4th ed.). Bradford, UK: Brill.

Quatieri, T. F. (2002). *Discrete-time speech signal processing: Principles and practice*. Bergen, NJ: Pearson Prentice Hall.

Chapter 3
History of Audio Forensics

The ability to perform audio forensic analysis hinges on the availability of audio recordings made outside the confines of a recording studio. The first portable recorders using magnetic tape appeared in the 1950s, and soon these devices were used to obtain clandestine recordings of interviews and wiretaps, as well as to record interrogations and confessions.

While the investigative usefulness of tape recordings was immediately apparent, the legal admissibility of recorded evidence was not immediately known. If the recording was obtained surreptitiously, would this violate the rights of the accused person against self-incrimination? What about uncertainty about the identity of voices and other details due to poor recording quality? What if the recording was a fake, or if it might have been altered or edited in some manner? Practical and legal concerns quickly became significant.

3.1 McKeever Case

Among the first notable legal cases involving forensic audio in the US federal courts was *United States v. McKeever* (United States District Court 1958). The case took place in the late 1950s. The government indicted the two defendants in the case, Thomas McKeever and Lawrence Morrison, under the federal anti-racketeering laws for committing and conspiring to commit extortion. McKeever and Morrison were agents for a local trade union branch of the International Longshoremen's Association, and the indictment charged the men with threats to a company holding contracts with the Longshoremen's union, James J. Ball & Sons.

In this case, the relevant audio recordings did *not* come from government surveillance, but from the defendant Thomas McKeever himself. After the indictments were issued, Mr. McKeever on his own arranged to go speak with representatives of the Ball Company while secretly arranging to have the conversations tape-recorded. McKeever's attorneys subsequently wanted to use the secret recordings to impeach

R. C. Maher, *Principles of Forensic Audio Analysis*, Modern Acoustics and Signal Processing, https://doi.org/10.1007/978-3-319-99453-6_3

the credibility of a prosecution witness, George Ball, by demonstrating that Ball's testimony in court was inconsistent with his prior statements, as recorded by the defendant.

During the defense cross-examination of Mr. Ball, the witness testified that he did not remember the particular conversation with Mr. McKeever that was of importance to the defense. The judge allowed the defense attorney to play a tape recording in open court for Mr. Ball to listen to via headphones, with the jury present but not hearing the recording itself. In other words, the judge allowed the recording to be used to revive or refresh Mr. Ball's memory, but not to be entered as evidence for the jury to examine. After hearing the tape, Mr. Ball testified that he did now recall the conversation and confirmed his prior testimony.

At that point, the defense argued that Mr. Ball's testimony in court was inconsistent with the tape-recorded conversation and requested that the court allow the jury to hear the part of the tape recording demonstrating the asserted inconsistency. However, the prosecution objected, arguing that the defense had not shown a solid foundation "for the accuracy or authenticity of the tape recording. There is no proof at this time before the Court that the tape recording produced by the defendants is a tape recording of any conversation had between McKeever and George Ball."

The court's decision on this question was to refuse to have the recording played in court, based on a review of several prior precedents regarding recorded evidence. The court stated (United States District Court 1958):

> A review of the authorities leads to the conclusion that, before a sound recording is admitted into evidence, a foundation must be established by showing the following facts:
>
> 1. That the recording device was capable of taking the conversation now offered in evidence.
> 2. That the operator of the device was competent to operate the device.
> 3. That the recording is authentic and correct.
> 4. That changes, additions, or deletions have not been made in the recording.
> 5. That the recording has been preserved in a manner that is shown to the court.
> 6. That the speakers are identified.
> 7. That the conversation elicited was made voluntarily and in good faith, without any kind of inducement.

The forensic audio community informally refers to these seven facts as the *Seven Tenets of Audio Authenticity*.

Although Tenets 1 and 2 seem obvious today, in 1958, there was likely more concern about the technical aspects of tape recording. Tenets 3 and 4 are still very important and relevant today: the court wants to be sure that there was no tampering with the recording, either accidentally or deliberately. Tenet 5 expresses the need for a verified chain of custody of the recording, just like any other physical evidence. Forensic examiners may be called upon to address Tenet 6, which states that the participants in the recording need to be identified. Finally, Tenet 7 gives the requirement that the recording was spontaneous and no one coerced the participants.

The court in the 1958 McKeever case also went on to say, presciently, that:

> Current advances in the technology of electronics and sound recordings make inevitable their increased use to obtain and preserve evidence possessing genuine probative value.

> Courts should deal with this class of evidence in a manner that will make available to litigants the benefits of this scientific development. Safeguards against fraud or other abuse are provided by judicial insistence that a proper foundation for such proof be laid

3.2 McMillan Case

In 1974, a federal narcotics conviction led to an appeal based on the use of audio recordings in the defendant's trial. A federal informant, Beverly Johnson, served as a go-between for heroin trafficking, and the federal agents monitoring Johnson used a recording device on her telephone to capture various conversations. Several conversations with suspect Harold McMillan involved arrangements for the purchase of heroin.

During McMillan's trial, the judge allowed the prosecutor to play excerpts of the recorded conversations for the jury, as well as allowing an agent to read the written transcript of the recordings. The defense objected to the use of the recordings and transcripts, arguing that the prosecutor had not established authenticity and legal foundation. As with McKeever, the appeal's court judgment reinforced the basic tenets of audio forensic admissibility and addressed some of the specific questions about establishing authenticity and talker identification.

3.3 FBI Procedures

The US Federal Bureau of Investigation (FBI) began performing audio forensic analyses and enhancements in the early 1960s. Expanding upon the McKeever tenets, the FBI established a 12-step procedure for processing audio recordings (Koenig 1988):

1. Evidence marking.
2. Physical inspection.
3. Recorded track position and configuration.
4. Azimuth alignment determination.
5. Playback speed analysis.
6. Proper playback setup.
7. Overall aural review.
8. Overall FFT review.
9. Setup of enhancement devices.
10. Copying process.
11. Work notes.
12. Reporting.

Steps 3–6 referred specifically to issues associated with analog magnetic tape recordings that were the only common recording medium available at the time.

3.4 The Watergate Tapes

In the wee hours of the morning on June 17, 1972, Frank Wills, a night security guard at The Watergate Hotel and office complex in Washington, D.C., discovered duct tape on the latch bolt of a basement access door. Mr. Wills thought that the tape had simply been applied and then forgotten during the day by construction workers to get in and out, so he removed the tape. Later during his rounds, he found that someone had reapplied the tape to the door latch, and Wills called the police. Security personnel soon found five burglars in the Democratic National Committee offices on the sixth floor. Little did anyone know at the time, the incident set off a series of events that would ultimately involve President Nixon's resignation—and also an important milestone in audio forensic analysis.

The Watergate burglars were eventually tried and found guilty of various federal charges, but by early in 1973, there was sufficient evidence to suggest that the burglary was a part of a larger set of questionable activities by the Nixon reelection campaign—perhaps with the support and direction of White House officials. The US Senate was sufficiently concerned that it formed the Senate Select Committee on Presidential Campaign Activities on February 7, 1973. Testimony soon pointed toward a larger conspiracy involving White House officials and the possibility that Nixon's advisors took steps to obstruct justice by covering up illegal conduct. Suspicions mounted during April, May, and June 1973, as the Senate committee continued its public hearings.

Then in July 1973, White House aide Alexander Butterfield testified before the Senate Committee and revealed for the first time the existence of audiotape recordings of conversations between the president and his advisors dating as far back as 1971. During Nixon's first term in office, the president directed the Secret Service to install audiotaping systems in the Oval Office and the Cabinet Room of the White House, in the president's private office in the Executive Office Building (EOB), and at Camp David in rural Maryland. Only the President and a small group of aides knew of the existence of these recording systems (Nixon Presidential Library and Museum 2015).

President Nixon initially refused to release the newly revealed "White House tapes," citing executive privilege, but transcripts and several specific tapes were subpoenaed late in 1973.

But already late in 1973, Watergate investigators became interested in one particular tape recording, a conversation between President Nixon and his Chief of Staff H.R. Haldeman. The investigators believed that the conversation, recorded in the Executive Office Building on June 20, 1972 (3 days after the Watergate break-in), likely included Nixon and Haldeman discussing the Watergate cover-up. However, the investigators found that the portion of the recording of interest to them was inaudible: the segment contained 18 ½ min of a buzzing sound, but no detectable conversation. The investigators suspected that the "18 ½-min gap" was evidence that someone had deliberately erased or recorded over the original conversation to destroy the incriminating portion of Nixon and Haldeman's conversation.

As the accusations grew, the possibility that someone had obstructed justice by deliberately erasing a portion of the tape came before John J. Sirica, Chief Judge of the US District Court for the District of Columbia. Judge Sirica determined in November 1973 that the potentially altered tape would need forensic study (McKnight and Weiss 1976). Watergate Special Prosecutor Leon Jaworski and James D. St. Clair, the counsel for the president, jointly nominated a group of six outside technical experts to form a special Advisory Panel on White House Tapes "… to study relevant aspects of the tape and the sounds recorded on it." The panel members were Richard H. Bolt, Franklin S. Cooper, James L. Flanagan, John G. (Jay) McKnight, Thomas G. Stockham, Jr., and Mark R. Weiss (Advisory Panel on White House Tapes 1974).

The experts on the Advisory Panel used a systematic analysis approach that is still considered the best practice for assessing audio authenticity. First, the panel examined the physical and mechanical aspects of the tape, looking for any signs of alteration or damage. Next, they documented to total length of the recording and attempted to verify that the recording was continuous and without unexplained erasures or start/stop sequences. They used critical listening of the entire recording and used nondestructive magnetic and electrical observation, plus signal processing for intelligibility enhancement.

The Panel reported in May 1974 that magnetic erasures caused the 18½-min gap at some point after the original recording session. The Panel identified several overlapping erasures performed with a specific model of tape recorder that differed from the device that produced the original recording. The panel's conclusion was based primarily on the characteristic start/stop magnetic signatures present on the June 20, 1972, tape.

Ultimately, on July 24, 1974, the US Supreme Court unanimously ordered Nixon to produce all of the relevant White House tape recordings. Among the recordings was a conversation from June 1972, a few days after the Watergate break-in, in which President Nixon agrees with a suggestion to direct CIA and FBI officials to involve themselves in the Watergate investigation on the grounds of national security concerns.

The revelations in this so-called "smoking gun" recording were widely viewed as a clear attempt by the President to obstruct justice. Thus, lacking any meaningful political support in congress, President Nixon resigned on August 8, 1974, rather than face the high probability of impeachment and removal from office.

3.5 Reevaluation of the Assassination of President Kennedy

On November 22, 1963, President John F. Kennedy was shot and killed in Dallas, Texas, while riding in a motorcade on Elm Street as his limousine passed through Dealey Plaza, just west of downtown. The gunfire also injured Texas Governor John Connally, seated in the limousine in front of the President. Gov. Connally ultimately recovered from his injuries. Few crimes have been subject to as much sustained interest, scrutiny, and speculation as the President's assassination.

The official finding of the Warren Commission investigation was that a single rifle was fired three times in succession by Lee Harvey Oswald from a sixth floor window of the Texas School Book Depository Building on Elm Street. One shot passed through the President's neck and then injured Governor Connally, one shot struck the President's head, and another shot that apparently missed the limousine entirely (Warren Commission Report 1964).

Unfortunately for investigators, there was very inconsistent earwitness testimony from Secret Service Agents, law enforcement officers, and bystanders who were in Dealey Plaza at the time of the shooting regarding the number of shots and the direction from which the shots came.

Abraham Zapruder, a civilian spectator on hand to see the motorcade, filmed an amateur 8 mm movie of the President's limousine as it made its way west on Elm Street in front of the School Book Depository Building. Amateur movie cameras in 1963 did not record sound, but the silent Zapruder film provided key evidence regarding the likely timing of the gunshots and the gruesome injuries inflicted upon the President and Governor Connally.

Despite the lack of audio accompanying the Zapruder film, investigators determined that Dallas Police radios might have picked up sound from Dealey Plaza at the time of the assassination. The Dallas Police Department used two radio channels for police dispatch communications on the day of the incident. One of the channels was used for routine radio traffic, while the second channel was used by officers involved in President Kennedy's motorcade. The audio from channel 1 was recorded using a machine known as a *dictabelt*, which employed a moving stylus to embed an analog groove in a flexible plastic belt moving through the machine. The resulting groove could be played back by running the grooved belt past a pickup stylus. The audio from channel 2 was also recorded, but the audio was embossed on a disc in a machine known as an *audograph*. Both the dictabelt and the audograph machines recorded in a voice-activated fashion to conserve recording time: the recorder stopped if the corresponding radio channel was silent, then started again when a radio message came through.

The Warren Commission examined the dictabelt and audograph recordings and transcribed the audible conversations. Then at some point following the formal investigation, it was asserted that a radio on a police motorcycle participating in the motorcade had somehow malfunctioned and had been transmitting continuously for a period of time. Although the regular motorcade dialog was on channel 2, the "open microphone" recording from the motorcycle appeared on channel 1, allegedly capturing the sound of the motorcycle engine and other background noises.

In 1978, the *House Select Committee on Assassinations* reopened several investigations into the John F. Kennedy assassination. Among the theories proposed was that the motorcycle with the open microphone could have been in Dealey Plaza, so the gunshot sounds might be detectable in the channel 1 dictabelt recording. The Committee hired Dr. James Bargar and a team from Bolt, Beranek, and Newman (BBN) to analyze a reference copy of the Dallas dictabelt. The BBN analysis included a series of test gunshots recorded at several locations in Dealey Plaza in order to reconstruct the circumstances of a rifle located in the Texas School Book Depository (as determined by the Warren Commission) as well as another firearm located east of the Book Depository in a park-like area known as the "grassy knoll." BBN concluded that the dictabelt recording did include three gunshot sounds attributable to a rifle located in the sixth floor window of the School Book Depository, but they also announced the stunning conclusion based on the dictabelt recording that there was very likely a fourth shot that they believed came from the grassy knoll area. This conclusion was remarkable, as it required a second gunman and presumably a previously unknown conspiracy!

The House Select Committee also hired Mark R. Weiss and Earnest Aschkenasy of Queens College, City University of New York, to perform an independent analysis of the dictabelt recording and BBN's assessment. Weiss and Aschkenasy came to the same conclusion as BBN, with an even higher stated probability of a shot from the grassy knoll (Weiss and Ashkenasy 1979). The spectacular acoustical findings became one of the key points in the Select Committee's final report.

However, other investigators and acousticians raised questions about the reliability of the acoustical evidence, the presumed location of the open microphone, the timing of the recording, and the methodology used to state the degree of scientific certainty of the findings. In 1980, the US Justice Department asked the National Academy of Sciences to perform another review of the acoustic evidence and the methods employed by BBN and by Weiss and Aschkenasy (National Academy of Sciences 1982). Also around this time, a private citizen named Steve Barber heard a publicly released copy of the channel 1 dictabelt recording and identified several issues with the recording. Most importantly, Mr. Barber noticed intelligible "cross talk" of utterances from some of the channel 2 conversations being present in the channel 1 recording. Specifically, the recognizable voice of Sherriff Bill Decker stating, "hold everything secure," is present in the dictabelt recording at approximately the same time as the purported gunshot sounds identified by BBN. The Sherriff's statement is known to have come about 1 min after the assassination, so the NAS report concluded that the dictabelt recording could not support the hypothesis of a second shooter on the grassy knoll, as the examination did not involve a portion of the timeline involving any gunshots.

Despite the multiple scientific examinations and various rebuttal reports, the arguments about the Dallas dictabelt evidence continue even to the present day.

3.6 Talker Identification and "Voiceprints"

The term *voiceprint* first appeared in Bell Telephone Laboratories publications as early as 1944 (Tosi et al. 1972). In 1962, Lawrence Kersta of Bell Labs published a paper in the journal *Nature* entitled "Voiceprint Identification" (Kersta 1962). The paper suggested that the individual dimensions of the talker's oral, pharyngeal, and nasal cavities could uniquely define speech spectrograms and that this could potentially allow a comparison between a recording of an unknown talker and a database of known recordings. If feasible, this appealing concept would be the aural equivalent of a fingerprint. Subsequent testing by Kersta and by Tosi et al. provided some promising results.

During the 1960s and 1970s, some audio forensics practitioners developed what became known as the *aural-spectrographic method* of comparing the spectrogram of an unknown talker with spectrograms from a set of known talkers. The method used a segment of recorded speech uttered by an unknown talker, such as from a telephone wiretap, answering machine, or surveillance system. The suspect then provided a segment of speech, often obtained using a script of words from the recording with the unknown talker. The examiner then used a combination of critical listening to the unknown and known talker recordings and visual comparison of the corresponding spectrograms to come to a conclusion about the likelihood that the suspect was the one who uttered the unknown recorded speech. The examiner reports one of five possible opinions:

1. Positive identification (the suspect's speech positively matches the unknown recorded speech).
2. Probable identification.
3. No decision.
4. Probable elimination.
5. Positive elimination.

Despite the appeal of the voiceprint concept, significant questions arose regarding the reliability and dependability of the aural-spectrographic technique for forensic applications. Other studies and reports eroded the underlying assumptions about the speech of an individual being spectrographically unique and time-invariant and called into question the likelihood of false identification or false elimination (Bolt et al. 1969, 1970, 1973).

In 1976, the FBI requested that the National Academy of Sciences appoint a special panel of the National Research Council to study the scientific principles and reliability of aural-spectrographic voice identification. The FBI noted that many court jurisdictions were seeing voice identification evidence, yet the controversies about admissibility and reliability remained in dispute. Ultimately, the special panel wrote (Bolt et al. 1979, p. 2):

> The Committee concludes that the technical uncertainties concerning the present practice of voice identification are so great as to require that forensic applications be approached with great caution. The Committee takes no position for or against the forensic use of the aural-visual method of voice identification, but recommends that if it is used in testimony, then the limitations of the method should be clearly and thoroughly explained to the fact finder, whether judge or jury.

Recent discussions of the aural-spectrographic method, such as Poza and Begault et al. (2005), still echo these caveats.

References

Advisory Panel on White House Tapes. (1974). *The executive office building tape of June 20, 1972: Report on a technical investigation*, United States District Court for the District of Columbia.

Begault, D. R., Brustad, B. M., and Stanley, A. M. (2005). Tape analysis and authentication using multi-track recorders, in *Proceedings of Audio Engineering Society 26th Conference, Audio forensics in the digital age, Denver, CO* (pp. 115–121).

Bolt, R. H., Cooper, F. S., David, E. E., Denes, P. B., Pickett, J. M., & Stevens, K. N. (1969). Identification of a speaker by speech spectrograms. *Science, 166*(3903), 338–342.

Bolt, R. H., Cooper, F. S., David, E. E., Denes, P. B., Pickett, J. M., & Stevens, K. N. (1970). Speaker identification by speech spectrograms: A scientist's view of its reliability for legal purposes. *Journal of the Acoustical Society of America, 47*(2), 597–612.

Bolt, R. H., Cooper, F. S., David, E. E., Denes, P. B., Pickett, J. M., & Stevens, K. N. (1973). Speaker identification by speech spectrograms: Some further observations. *Journal of the Acoustical Society of America, 54*(2), 531–534.

Bolt, R. H., Cooper, F. S., Green, D. M., Hamlet, S. L., McKnight, J. G., Pickett, J. M., Tosi, O. I., & Underwood, B. D. (1979). *On the theory and practice of voice identification*. Washington, DC: National Academy of Sciences.

Kersta, L. (1962). Voice print identification. *Nature, 196*, 1253–1257.

Koenig, B. E. (1988). Enhancement of forensic audio recordings. *The Journal of the Audio Engineering Society, 36*(11), 884–894.

McKnight, J. G., & Weiss, M. R. (1976). Flutter analysis for identifying tape recorders. *The Journal of the Audio Engineering Society, 24*, 728–734.

National Academy of Sciences. (1982). *Report of the committee on ballistic acoustics*. Washington, DC: National Academy Press.

Nixon Presidential Library and Museum. (2015). *History of the White House tapes*.

Tosi, O., Oyer, H., Lashbrook, W., Pedrey, C., Nicol, J., & Nash, E. (1972). Experiment on voice identification. *The Journal of the Audio Engineering Society, 51*(6), 2030–2043.

United States District Court, Southern District, New York. (1958). *U.S. v. McKeever, 169 F.* Supp. 426 (S.D.N.Y. 1958).

Warren Commission Report. (1964). *Report of the President's commission on the assassination of President John F. Kennedy*, U.S. Government Publishing Office, https://www.gpo.gov/fdsys/pkg/GPO-WARRENCOMMISSIONREPORT/content-detail.html

Weiss, M. R., & Ashkenasy, E. (1979). *An analysis of recorded sounds relating to the assassination of President John F. Kennedy*, U.S. Congress, House of Representatives, House Select Committee on Assassinations Proceedings, vol. 8.

Chapter 4
Handling Forensic Evidence

An audio forensic investigation involves handling evidence. Often the evidence may be simply a digital file transferred from a compact disc, a USB memory stick, or even an email attachment. In other cases, the evidence may be stored in some sort of proprietary manner internally in a device or surveillance system. Even today, some audio forensic evidence may come as an analog tape recording.

An official forensics lab will have standard practices and procedures for processing audio evidence and standard training procedures for all of the examiners. These procedures generally can be expected to follow the same principles applied to handling other types of physical evidence. If the organization does not have its own guidelines, example guidelines such as the best practices promulgated by the Scientific Working Group on Digital Evidence (SWGDE) should be a good starting point.

4.1 Basic Tools: Audio Playback, Waveform View, and Spectrographic View

The fundamental tools of contemporary audio forensic examination are a good-quality audio playback system, a waveform display program, and a spectrographic display program. These functions are typically performed using a conventional desktop or laptop computer.

4.1.1 Audio Playback System

The audio playback system needs to be of sufficient quality and flexibility that exceeds the frequency content and dynamic range of the forensic audio material. In other words, any quality limitations will be attributable to the audio recording, not to problems with the playback system.

R. C. Maher, *Principles of Forensic Audio Analysis*, Modern Acoustics and Signal Processing, https://doi.org/10.1007/978-3-319-99453-6_4

The computer's built-in audio subsystem, soundcard, or USB-attached converter must support an appropriate range of sampling rates and formats, as well as the various audio format decoding and reconstruction software modules needed to deal with the native format of the audio forensic evidence. Satisfactory loudspeakers are generally available from reputable manufacturers supplying professional general purpose recording studio monitors. A good guideline for stated frequency response would be 50 Hz and 20 kHz. Headphones are recommended for many audio forensic tasks, as they tend to reduce the effects of room reverberation, computer fan noise, and other audible distractions in the playback environment. Look for professional-quality headphones with comfortable earpieces that completely seal around the ears, and arrange the playback system so that the headphone system has a separate volume control knob.

While there may be a tendency to want to turn up the sound level when trying to hear potentially relevant sounds in a low-quality audio forensic recording, it is important to avoid making the level so loud that it causes the ear to adapt with reduced sensitivity (the acoustic reflex). It is also important to listen to the recording in such a way that unexpected loud sounds will not hurt the ears.

4.1.2 Waveform View

Interpretation of aural information requires the ears, but the *eyes* can also be helpful in audio forensic analysis. The fundamental visual task is graphical waveform display, which depicts the audio recording as a graph with time on the horizontal axis (abscissa) and amplitude on the vertical axis (ordinate). Waveform display programs generally allow viewing a specific time range, with controls to "zoom in" or "zoom out" the time axis and the amplitude axis.

The graphical display usually shows the individual waveform samples as dots if the time interval is very short. Some display programs have a "connect-the-dots" display that draws lines between the individual sample points, as shown in Fig. 4.1. If the time interval is made longer, there may be more samples to display than there are horizontal pixels on the display screen, and most display programs will show the maximum and minimum sample amplitudes in a short time span, creating the *envelope* of the audio signal, as depicted in Fig. 4.2.

The most useful graphical waveform display programs also provide simultaneous audio playback: a pair of cursors (or select-and-drag highlighting) identifies the chosen playback start and stop positions in the waveform. This allows an iterative procedure of listening and watching as the waveform details change aurally and visually.

Display programs often include waveform editing features, storage format conversion, audio effects processing, and many other useful features. However, it is important to guard and maintain the original reference copy of the audio, and in particular, prevent inadvertent edits during the viewing and initial assessment procedures.

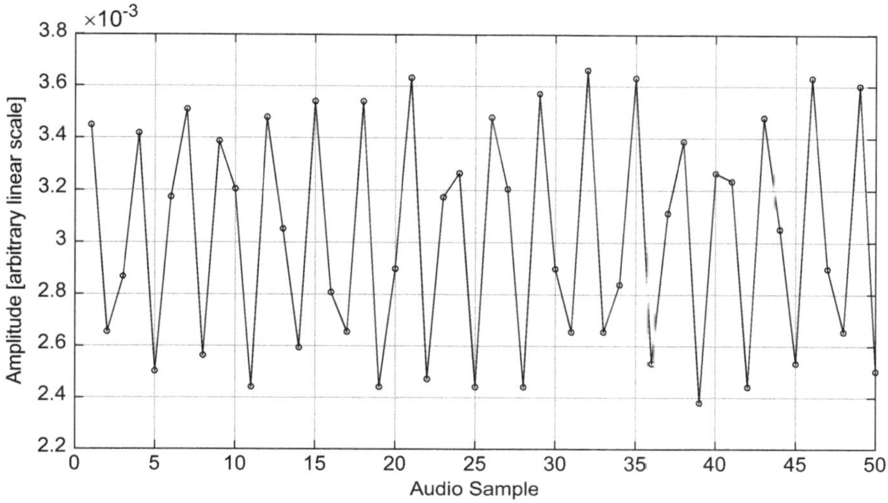

Fig. 4.1 Digital audio display of a time span sufficiently short to show individual samples, with "connect-the-dots" lines between the sample points

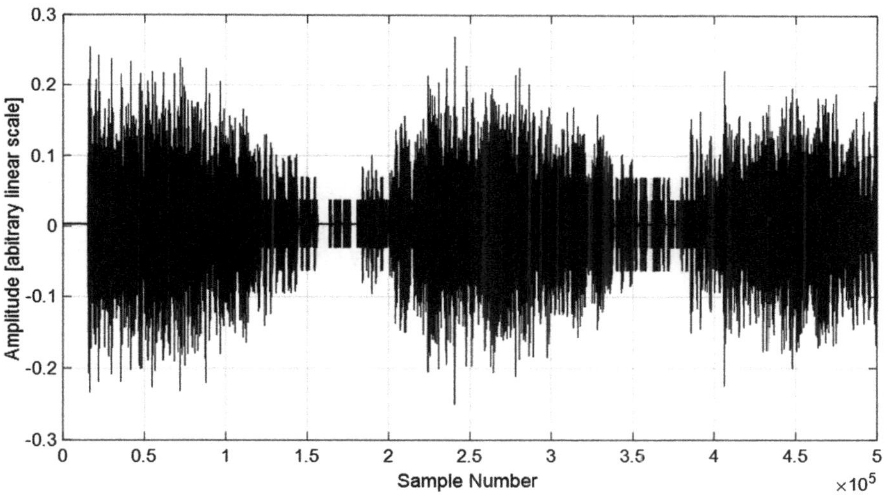

Fig. 4.2 Waveform display of a time span that is too long to depict every individual sample: the display shows the signal envelope

One particular concern occurs when working with encoded audio files, such as MP3. In order to view and to listen to the file, the display program decodes the MP3 file into regular pulse-code modulation (PCM) samples. If the file is edited in any way and then saved again as MP3, the PCM samples are *re-encoded* as MP3, creating a second generation of encoding. Because MP3 and other similar perceptual coders are *lossy*, the decode/re-encode/decode cycle tends to create a build-up of audible distortion with each lossy encoding step. As noted in the previous chapter, it is important not to decode, alter, and then re-save the edited file in encoded format.

4.1.3 Spectrographic View

A very useful method for visual display of audio forensic recordings is the *spectrogram*. The spectrogram is a special type of graph produced by calculating the short-time Fourier transform magnitude (the *spectrum*) of successive brief time intervals of the input signal and displaying them sequentially across the screen (Allen and Rabiner 1977). The spectrogram takes successive short blocks, or frames, from the audio signal recording, as shown in Fig. 4.3.

Like the waveform display, the spectrogram presents a graph of audio signal energy with the horizontal axis being the time scale. Unlike the waveform display, the vertical axis of the spectrogram is the signal *frequency* scale in hertz. The relative amount of audio signal energy at a particular time and a particular frequency is given by the color or brightness of the spectrogram at the corresponding time and frequency coordinates in the graph. For this reason, the spectrogram is sometimes referred to as showing the signal in the *frequency domain*, while the waveform display shows the signal in the *time domain*, as shown in Fig. 4.4. The upper

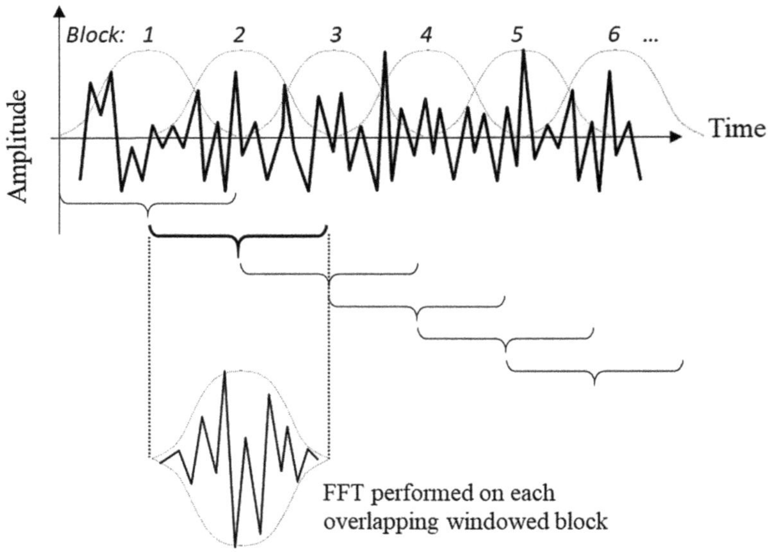

Fig. 4.3 Short-time Fourier transform concept. The audio signal is segmented into overlapping blocks or frames, and the fast Fourier transform (FFT) is used to calculate the short-time spectral magnitude of each block

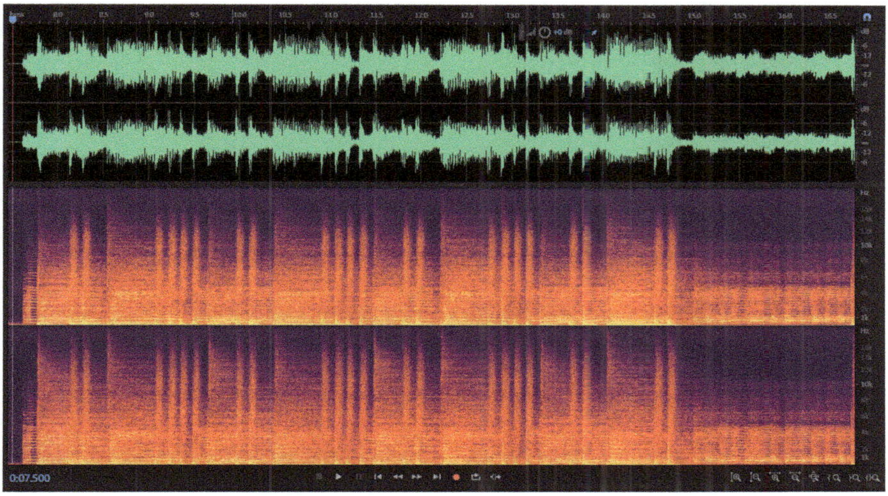

Fig. 4.4 Combined time domain and spectrographic display of a stereo (2-channel) audio recording of a rock-and-roll instrumental combo (electric guitar, bass, and drums). The overall duration is 10 s, with the time scales the same for each of the four displayed panels. The frequency range in the lower two panels (vertical axis) is 0–20 kHz, log scale. The upper two panels (light green hue) display the time domain envelope (signal amplitude vs. time) of the audio signal for the left channel (top row) and right channel (second row). The lower two panels (orange hue) show the left channel and right channel spectrograms, respectively. The spectrogram panels have frequency 0–20 kHz on the vertical scale and time on the horizontal scale. The spectral energy is depicted by the brightness of the colors in the spectrograms: less energy has dark-colored pixels at the corresponding time and frequency, while more energy has bright-colored pixels at the corresponding time and frequency. Note the repeated pattern of vertical reddish bars in the spectrogram due to drum hits and the horizontal yellow stripes at lower frequencies due to harmonics of the electric guitar and bass lines

portion of the display shows the time waveform envelope for the two stereo channels, while the lower portion of the display shows the spectrogram of each channel.

In the spectrographic view, an impulsive sound such as a click or gunshot appears as a vertical line, indicating that there is energy across frequency (broad along the vertical axis) but that it lasts for a brief instant (short along the horizontal axis). Conversely, a whistle or continuous hum tone appears as a horizontal line, indicating that the sound energy is relatively discrete in its frequency extent but continuous in duration (Fig. 4.5).

Spectrographic display programs allow viewing a specific time range and usually allow specification of the frequency range. However, it is important to understand that there is a fundamental mathematical trade-off between signal resolution in time and in frequency. Zooming in on a very short-time duration of a signal inherently prevents simultaneously fine frequency resolution, while zooming out for a longer time duration allows finer resolution of frequency detail, but the longer time observation "window" prevents knowing details about when in time a particular signal occurred. In other words, the spectrogram has a trade-off between how selective the

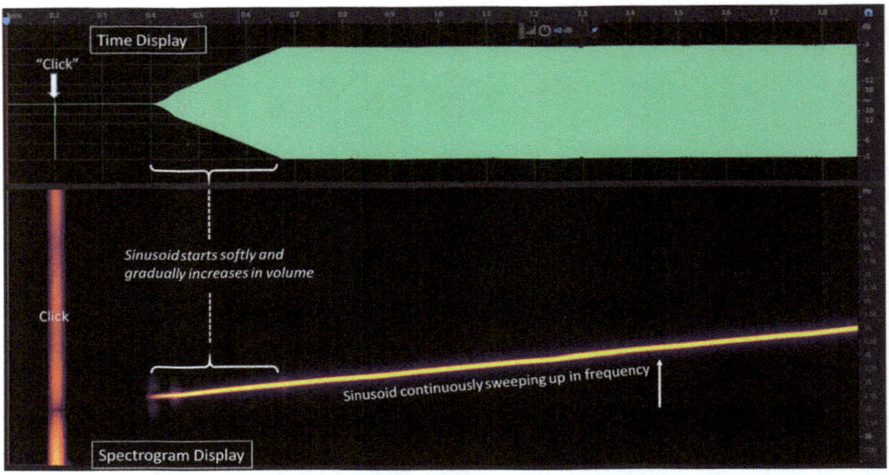

Fig. 4.5 Spectrogram of a "click" followed by a "tone" sweeping up in frequency (overall duration 2 s, frequency range 0–10 kHz, linear scale)

display can be in separating signal components of similar frequency and how detailed the timing can be. This trade-off is shown in Fig. 4.6.

4.2 Starting the Examination

Among the challenges of forensic examination is avoiding *bias* in the interpretation process. In the context of audio forensic examination, bias often comes from the use of extraneous non-audio information about a case, the suspects, the circumstances, and the investigator's suspicions. For example, the individual requesting the audio forensic examination may want to talk about the arrest history of a suspect, describe the physical evidence collected at the crime scene, suggest the desired conclusion that would "help" build the case, or divulge potentially incriminating comments by various individuals involved in the incident. While these details may be interesting and ultimately useful to the court or to a jury, the statements can also be prejudicial to the audio forensic examination process. The provided information, not drawn from the audio evidence, may influence the forensic examiner's work, either consciously or unconsciously.

As noted previously, the role of the forensic examiner is to educate the court about nature and reliability of the audio evidence from a scientific standpoint. *The examiner is not an advocate for a particular side in the adversarial legal process, but an expert who testifies solely regarding the audio evidence presented.* The audio forensic examiner's testimony addresses the facts, methods, and interpretation of the audio evidence. It is then up to the law enforcement investigators and the

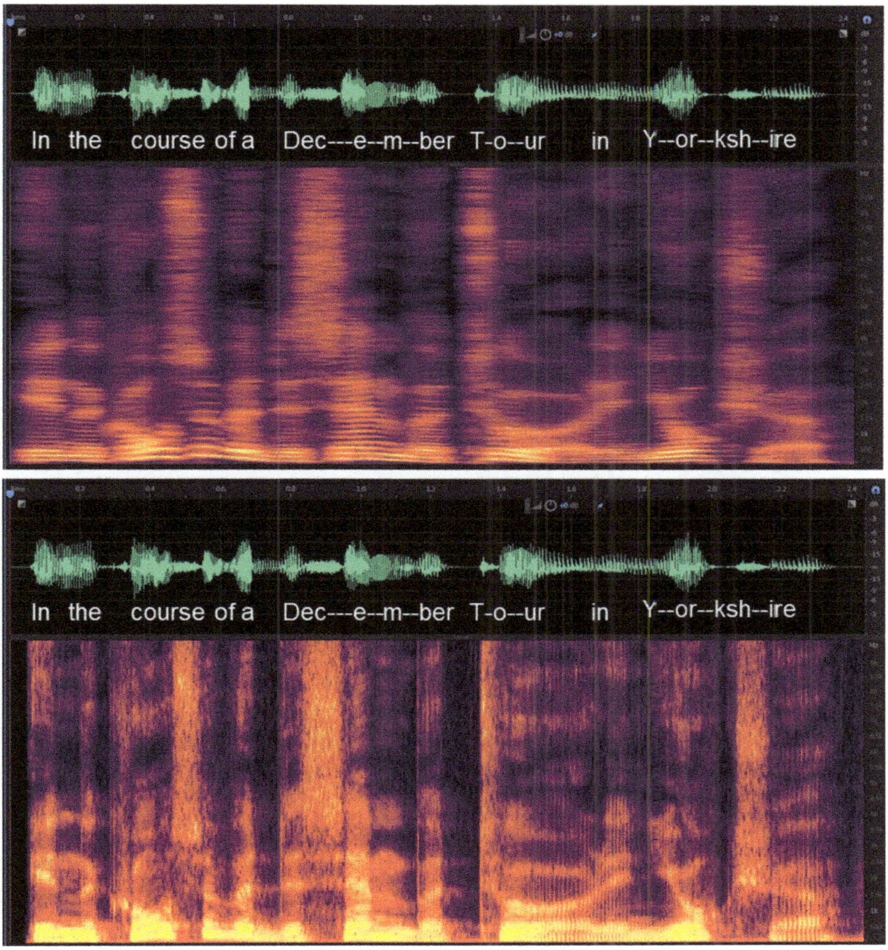

Fig. 4.6 Two spectrograms of the same speech utterance by a male talker showing the fundamental trade-off between time and frequency resolution. Upper frame: longer time block lengths give better resolution in frequency to show detail of harmonic partials, but this blurs the sound's attack and releases depiction in the spectrogram. Lower frame: shorter time block lengths provide better resolution in time to show "edges" when signal changes occur, but this blurs the frequency detail. Overall duration 2.5 s, frequency range 0–10 kHz, linear scale

attorneys to combine various pieces of evidence in a manner that will further their theories of the case.

Audio forensic examinations typically start with an inquiry from a law enforcement organization or an attorney. The requester may or may not be familiar with audio forensics procedures, so it is helpful to have a checklist, such as:

- Is the original audio recording available? If not, what is the nature of the best-available duplicate recording?

- Under what circumstances was the recording made?
- Is the recording of good quality, marginal quality, or poor quality?
- Is there a dispute about any aspect of the recording, such as its authenticity?
- Has any prior audio forensic examination been conducted? If so, what is the reason for the additional requested analysis?
- What are the specific audio forensic questions to be addressed?

Most audio forensic analysis requests require specific training and experience. It is imperative that the examiner declines engagements that go beyond his or her knowledge level.

It is vital to keep complete notes and documentation of all forensic engagements. Notes should be sufficiently detailed that the requests and processes can be recalled months or even years later. It is good practice to prepare notes and documentation in such a way that another examiner could read the description and have a very good idea of the audio forensic processes and conclusions.

It is highly recommended to start with the original recorded media and, if possible, the original recording system and create verified digital work copies before commencing any enhancement or interpretation work. The original recording system may allow retrieval of special device settings, proprietary native data, time-stamps, metadata, and other recording settings. If the device has special cables, power supply, connectors, etc., these also need to be requested.

Certain recording devices may have volatile memory: the recorded signal is lost if power is lost (e.g., batteries run down). Care must be taken to ensure that the memory is protected from potential power loss.

The audio forensic examiner should ask the sender to secure the evidence with "write protection" tabs and any other mechanical overwrite prevention settings.

The standard procedure for audio forensic examination in formal laboratory protocols gives the required evidence to accompany the investigation. The individual or agency providing the audio evidence needs to follow the protocol expectations (Scientific Working Group on Digital Evidence 2008). The requested information may include:

- The original recording or an exact digital duplicate copy.
- The equipment used to make the recording or a complete list of components, models, and serial numbers. The user manuals and any other descriptive material should also be available.
- All records of maintenance or repair to the recording equipment.
- Details of the recording method and circumstances, including the location, background sound level, power source for the recorder, the identity of all parties recorded, and details of the foreground and background sound sources (speech, music, radio, unrelated conversations, etc.).
- All details regarding the recording process, such as the number of times the recorder was stopped and started, changes to the recording level, use of voice-activated recording features, and so forth.
- Any available prior reports, transcripts, investigator notes, etc.

Once the audio forensic evidence/equipment is received, the audio forensics examiner will need to follow the laboratory standard practices (Audio Engineering Society 1996). These practices generally include:

- Maintaining the chain of custody—Record the date and circumstances under which the evidence was received, and ensure that the evidence is secured while under review to prevent damage or loss.
- Observation of the data carrier, metadata, and other details—Use photographs and written notes to document all of the materials submitted and how it was packed, including model numbers, serial numbers, formats, etc. Make particular note of any cracks, marks, scratches, or other damage.
- Initial/label nondestructively—Follow the laboratory policy regarding how the evidence should be uniquely marked so that it can later be distinguished from other evidence. Some labs will use a case number and date marking, while others will simply have the examiner's initials and date on the item. Use particular care if marking CD/DVD material and similar data carriers so that the markings will not damage the media. If it is not safe or if it is physically impossible to mark the item, place the data carrier in a suitable sealable container and mark the container (initial, date).
- Work with a verified digital copy, *never the original*, unless absolutely necessary. With analog evidence, a high-quality digital copy is made from the analog original. This may entail finding the proper playback equipment, aligning it to match the tape, and ensuring that the tape is of sufficient integrity that it can be played without causing damage. In such cases, it is recommended to seek the help of an analog specialist.

With digital audio evidence, direct digital "bitstream" copies should be made and verified. Care must be taken that the copying operation does nothing to alter the original contents. Many digital forensics laboratories use a hardware write blocker device between the storage device and the control computer. The write blocker intercepts any commands that would modify the storage contents so that the material is unaltered.

4.2.1 Initial Aural Evaluation

The first step in an audio forensic evaluation is to listen to the verified work copy of the audio material. Use a quiet environment with the sound playback level at a comfortable setting for this initial listening. Initial listening with loudspeakers is satisfactory if the playback area is free of distractions. It is standard practice to make preliminary notes about the audio material during this initial overall aural review and include any initial impressions about the quality and any noticeable defects or audible events in the recording.

Many forensic examiners will also choose to view successive spectrograms of the recording, using suitable time and frequency ranges. The spectrogram can often help identify subtle aspects of the signal and any background sounds in the recording for additional evaluation.

Following the initial listening and spectrographic observation, the examiner will then turn to the audio forensic questions posed by the requesting party. The minimum suite of analysis procedures includes *critical listening*, *waveform analysis*, and *spectral analysis*.

4.2.2 Critical Listening

As its name implies, *critical listening* is careful, focused audition of the forensic recording. Critical listening sessions must be in very quiet surroundings, free of distractions and interruptions, and generally require listening with comfortable high-quality headphones. As noted previously, the playback level is kept moderate to prevent aural fatigue and to avoid triggering the acoustic reflex (lowered sensitivity). The critical listening process should be *iterative*, meaning that after listening to the entire recording, the examiner "rewinds" to re-listen to important sections several times in succession. Many examiners choose to perform critical listening using a waveform display program, since the software makes it easy to place time markers and other annotations.

An important aspect of critical listening is to focus attention deliberately on the *foreground* sounds, such as speech dialog, and then during subsequent replays, focus deliberately upon the *background* sounds, such as ambient environmental noise, distant conversations, and subtle rattles. In certain circumstances, the background sounds may help identify the place and time of the recording, and in other circumstances, irregularities in the background sounds may be a clue to an edited or otherwise altered recording.

Repetitively listening to a short loop segment of the audio may seem like a way to glean subtle details, but the examiner has to be careful to avoid creating the mental impression of a percept that is based on the looping rhythm rather than the audio evidence itself.

4.2.3 Waveform Analysis

The ears can be extremely adept at detecting and identifying sounds, but not so adept at measuring precise time instants and amplitudes. The waveform display program provides a visual depiction of the audio signal, and this display can help identify audible events, time intervals, signal changes, and other signal attributes.

It is common for a forensic examiner to use the waveform display initially with a broad time range, perhaps as much as a few minutes, to get an overall impression of the signal waveform. Then the strategy is to zoom in successively on time intervals of interest, taking notes and making preliminary observations about the signal. Any of the signal contents that are of interest to the investigation, such as the time of a particular utterance or a distinctive background sound, get special scrutiny at this point. A good approach is to use an alternating combination of identifying signal features visually and listening to the signal aurally.

In the zoom-in mode, the examiner should look carefully for discontinuities, dropouts, abrupt clicks, and similar waveform irregularities that could indicate a problem with the recording system or the possibility of a deliberate deletion or alteration of the material.

4.2.4 Spectral Analysis

In addition to the waveform time domain display, viewing the spectrogram can help identify signal features of interest. With some practice, one can pick out important signal characteristics and changes from the spectrogram and then go back and listen to the audio signal corresponding to the spectral feature.

Recognizing the spectrogram's inherent time-frequency trade-off, the examiner may choose to switch among several different settings for frequency and time resolution. Reducing the analysis block length gives a better indication of *when* a sonic event occurred in the spectrographic display, while increasing the block length gives more resolution in the frequency dimension, but reduces the time resolution, blurring out the beginning and the end of the sound event, as shown in Fig. 4.7.

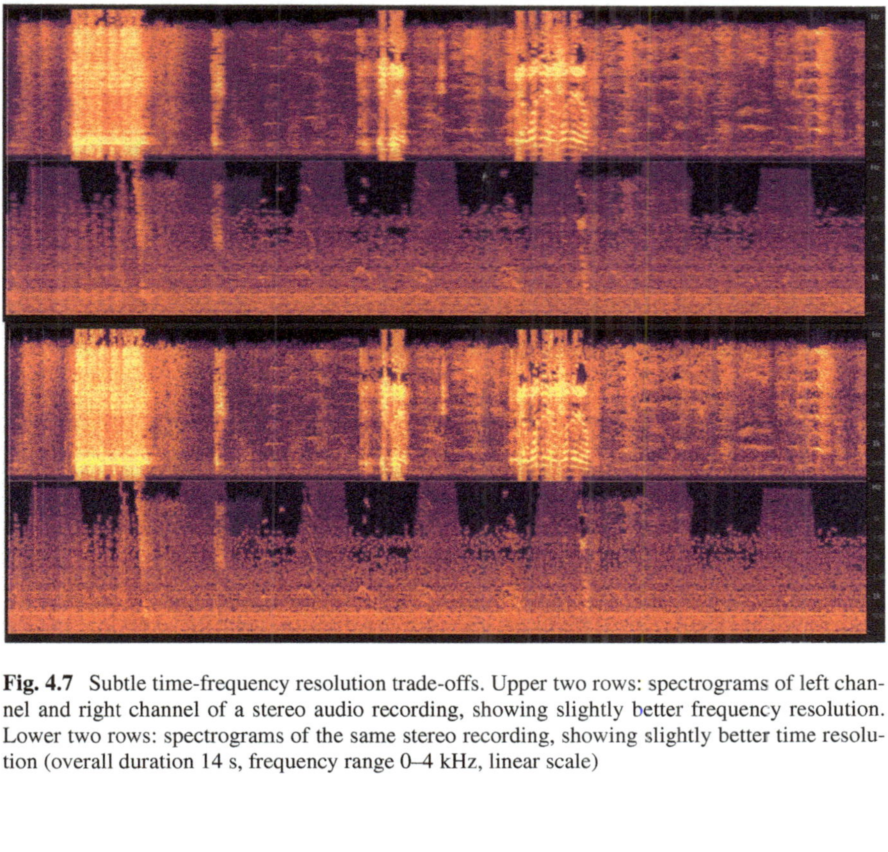

Fig. 4.7 Subtle time-frequency resolution trade-offs. Upper two rows: spectrograms of left channel and right channel of a stereo audio recording, showing slightly better frequency resolution. Lower two rows: spectrograms of the same stereo recording, showing slightly better time resolution (overall duration 14 s, frequency range 0–4 kHz, linear scale)

Along with the basic time-frequency trade-off selection, another common user option with spectrographic display software is the choice of *window* function. This refers to the use of an amplitude *weighting* that smoothly fades in and fades out the short-time block of audio used for each spectrographic segment, thereby avoiding some of the undesirable spectral effects of abruptly starting and stopping the data block. Common amplitude window functions in digital signal processing have been given nicknames, such as triangular, Bartlett, Hann, Hamming, Kaiser, Blackman-Harris, and so forth. If no tapering is used, the implicit window is referred to as "rectangular."

While the amplitude window mitigates the abrupt boundaries of each block, the window also has the side effect of reducing the spectral resolution to some extent. The precise shape of the amplitude window function has subtle effects on the frequency resolution, so it may be useful to experiment with different window functions, block lengths, and so forth, to help visualize the spectrographic details of greatest interest in a particular investigation.

Some display programs include simultaneous presentation of the time waveform, spectrogram, and audio playback, as was shown in Fig. 4.4. This allows a very flexible system for critical listening and visual assessment of signal characteristics, and this capability is highly recommended.

As previously noted, keep complete and comprehensive work notes during the aural/visual assessment. It is very common to have weeks or months between the initial observation of the evidence and subsequent steps, such as report writing and testimony. Details that may seem obvious on the first examination need to be written down for future use, not simply committed to memory.

References

Allen, J. B., & Rabiner, L. R. (1977). A unified approach to short time Fourier analysis and synthesis. *Proceedings of the IEEE, 65*(11), 1558–1564.

Audio Engineering Society. (1996). *AES27-1996: AES recommended practice for forensic purposes – Managing recorded audio materials intended for examination*. New York: AES.

Scientific Working Group on Digital Evidence. (2008). *SWGDE best practices for forensic audio* (Version 1.0).

Chapter 5
Authenticity Assessment

In certain cases, there may be a question about the *authenticity* of an audio forensic recording. Like any physical evidence, audio forensic recordings are subject to potential questions about authenticity: is the recording complete, unaltered, and consistent with the stated circumstances of its creation? For example, an individual may claim that a recorded conversation has been edited so that certain critical utterances are inserted or edited out. Other cases may involve suspicion that the asserted time, place, and circumstances are not what was claimed (Audio Engineering Society 2000). What is authenticity? Can it be guaranteed?

Recordings are always susceptible to accidental alterations or deliberate tampering, and detecting these changes may or may not be possible. The court must be convinced of the authenticity and integrity of the audio evidence. Audio forensic examiners must follow chain-of-custody procedures and avoid any possibility of unintended changes to the original evidence and must be diligent about potential signs of alteration. The court must also understand that the fact that an examiner does not find specific evidence of tampering does NOT necessarily mean that the audio recording is authentic: a particularly skilled adversary could conceivably create a tampered recording that defies detection.

5.1 Historic Context: Authenticity of Analog Magnetic Tape Recordings

Until the first decade of the twenty-first century, the primary medium for audio forensic evidence was analog magnetic tape. With the exception of a few mechanical recording systems, such as the dictabelt system used in the Dallas Police Department at the time of the John F. Kennedy assassination, magnetic tape was essentially ubiquitous as the means to capture live audio.

R. C. Maher, *Principles of Forensic Audio Analysis*, Modern Acoustics and Signal Processing, https://doi.org/10.1007/978-3-319-99453-6_5

Magnetic tape consists of a thin, flexible plastic ribbon that serves as a substrate for a thin layer of magnetic powder material impregnated in a binder substance and spread uniformly onto one side of the tape. A magnetic field deliberately brought into the vicinity of the tape can magnetize the surface material, leaving a telltale magnetic polarization. Later on, a magnetic detector circuit can measure the amount of magnetization of that particular segment of the tape.

The magnetic tape is stored rolled onto a spool, known as a reel. The magnetic tape recorder draws the tape off the supply reel at a fixed rate using a motor-driven capstan spindle and pinch roller. The tape path slides over three electromagnetic coils: the *erase head*, the *record head*, and the *playback head* (some less expensive tape recorders used just two heads: the erase head and a combined record/reproduce head). When recording, the moving tape first passes over the erase head to randomize the magnetic domains on the tape and then continues and passes over the small coil of wire in the record head that acts as a variable electromagnet. The electrical current through the record head's electromagnet is modulated by the analog audio signal, causing fluctuating magnetization of the tape to represent the audio information. The tape transport then collects the recorded tape and spools it onto a separate take-up reel.

To playback the recorded information, the tape is first rewound from the take-up reel back onto the supply reel. Next, the tape is passed again from the supply reel to the take-up reel, but the erase head and record head are not activated, while the playback head detects the magnetic field on the tape and regenerates the analog audio signal. Once the tape has been recorded, it can be played back repeatedly at will, with only gradual losses due to mechanical wear as the tape is moved through the player.

The relationship between the record head current and the magnetization of the tape is nonlinear, which causes distortion. To minimize this inherent distortion, a strong, inaudibly high-frequency (e.g., 40 kHz) AC *bias* signal is added (mixed) with the audio signal. The bias signal causes the tape to be strongly magnetized at the bias frequency, with a small residual magnetization being the audio signal component. The playback system reproduces only the audio frequency range, so the ultrasonic AC bias linearizes the overall behavior without interfering with the audible information.

Audio tape recorders can have multiple parallel magnetization heads to create multiple longitudinal *tracks* on a single tape. Multi-track recordings for consumer products typically have two tracks, corresponding to the left and right stereo channels. Common consumer devices, such as compact cassette recorders, also allow interleaved tracks: one stereo pair is recorded on two tracks of the tape, then the tape can be flipped over, and a second pair of left and right tracks recorded with the tape moving in the opposite direction. A few common tape track configurations are shown in Fig. 5.1.

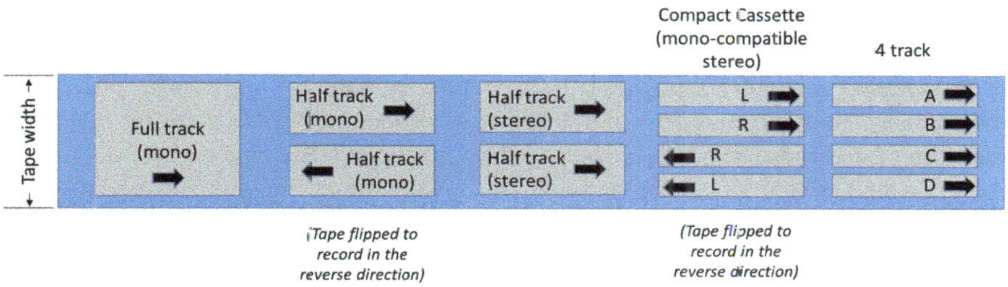

Fig. 5.1 Track format examples for analog audio magnetic tape

5.1.1 Physical Inspection

Audio forensic examination to assess the authenticity of an analog magnetic tape recording requires physical handling and examination of the tape itself (Audio Engineering Society 2000). Analog tape alterations made by physically cutting and then reattaching the tape, known as *splice* edits, involve adhesive tape used to hold the ends of the cut tape together.

The examiner will visually inspect the cassette housing, the reels, the entire length of the tape, and any related material, looking for spliced tape, broken housing, or other indications that the tape has been altered physically. The examiner will record any manufacturing serial numbers and tape batch designations, determining if the age of the tape is at least as old as the date the recording was reportedly made.

If the recorder used to produce the recording is available, the examiner inspects and tests the device. A qualified tape recorder technician can examine the track configuration, head alignment, azimuth setting, bias level, and so forth. If the recorder was out of calibration, it may be necessary to set up the playback head so that it matches the tape's alignment.

5.1.2 Magnetic Development

Authenticity evaluation of analog magnetic tape typically requires *magnetic development* to view the latent magnetic domains recorded on the tape. Magnetic development uses a ferromagnetic fluid (ferrofluid), which contains microscopic magnetic particles suspended in a solvent mixed with a surfactant to help keep the particles dispersed and suspended. The examiner spreads the ferrofluid uniformly, but sparingly, on the magnetic tape, which allows the suspended ferro particles to align with the invisible magnetic domains recorded on the tape. After allowing the solvent to evaporate, the examiner uses a microscope to observe the pattern of the magnetic particles adhering to the tape, known as *Bitter Patterns*, after Francis Bitter (1902–1967), a researcher at Westinghouse Electric Company and later MIT, who proposed the powder pattern method in 1931.

The erasing and recording process of analog tape creates a distinctive magnetic pattern when the recorder is started and stopped. The magnetic heads are energized as the capstan and reel motors start transporting the tape through the recorder, and the transient start-up magnetic fields leave a corresponding trace in the magnetic recording tape. Similarly, when the recording is stopped, the tape comes to a halt as the erase and record heads are de-energized. An example magnetization pattern is shown in Fig. 5.2. The image shows two regions of recorded material on a piece of analog cassette tape, caused by a stop/start recording sequence. The portion on the right is magnetization from the recording process up until the recorder was stopped, leaving the unmagnetized (dark) gap. Then the recorder was started again, which caused a slight offset of the magnetic pattern as the tape started moving again. The vertical striations in the magnetic patterns are due to the high-frequency AC bias mentioned previously (Koenig 1990).

The audio forensic examiner looks for the distinctive erase and record head magnetic signature patterns on the tape, as well as the magnetic tracks containing the audio information. If the recording is authentic, the examiner expects that there is a single start-up transient at the beginning of the recording, and then no other head transients until the recording was stopped. Any observation of additional start-stop sequences or erasures could indicate that the tape has been altered, deliberately or inadvertently, and the investigators would need to seek an explanation for why the recording appeared to be edited or truncated.

As was noted previously in the Watergate tape study, the investigators identified several overlapping erasures performed with a specific model of tape recorder that differed from the device that produced the original recording based on the characteristic start/stop magnetic signatures present on the June 20, 1972 tape (see Fig. 5.3).

In recent years, the use of a multi-track tape recorder to read information from the original track format on the tape has been used (Begault et al. 2005). Also,

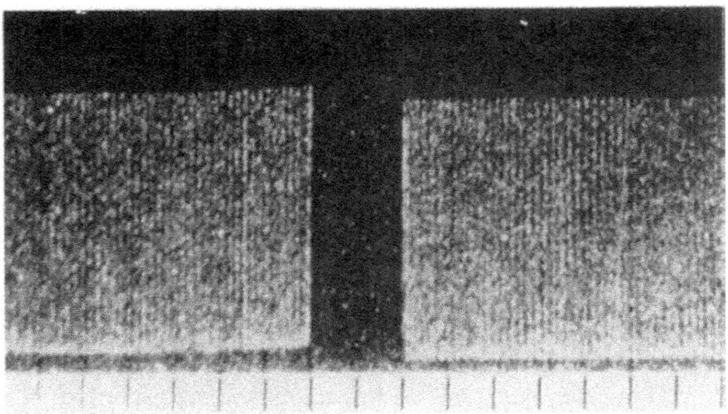

Fig. 5.2 Example magnetic domain "Bitter Patterns" magnified from an analog audio tape recording. The dark gap in the middle of the photograph shows where a recording had been stopped and then re-started (from Koenig 1990, reprinted by permission)

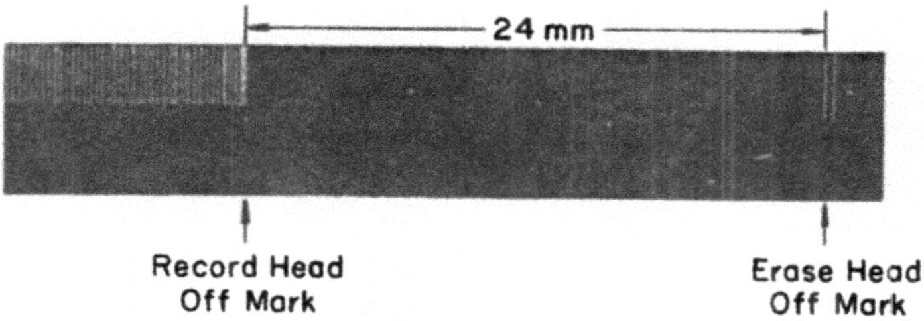

Fig. 5.3 Example of magnetic development as used in the Watergate tape investigation report (from Advisory Panel on White House Tapes 1974)

several high-resolution direct imaging methods have been developed using specialized equipment to reveal the recorded magnetic pattern without the use of the ferrofluid (Marr and Pappas 2008).

Some authenticity issues with analog tape recordings may be due to an edited tape that is subsequently copied and then presented as if it were an original recording. The copy can even be made with a single start and stop record sequence, so it may appear to be a continuous, authentic recording. In such a case, there may still be other evidence of tampering detectable by signal irregularities or gaps, as described below.

5.2 Current Context: Authenticity of Digital Audio Recordings

A digital audio recording presents many challenges for authenticity assessment (Brixen 2007). Digital audio recordings are essentially sequential lists of binary numbers stored in a digital computer file, and digital files can be copied, transmitted, and stored on a variety of media with perfect fidelity. What's more, it is often difficult to exclude the possibility that a digital file was adjusted and edited surreptitiously and then stored as a seemingly intact and pristine file. If all that is available is the digital audio file itself, the examiner must use other means to assess the integrity of the recording.

5.2.1 Identifying Edits: Splicing and Mixing

An audio forgery could consist of one or more edits made to an original recording by deleting certain time segments, by inserting audio material, or by additively mixing in the forged material. An unsophisticated forger could attempt to make such edits in the digital audio file with an abrupt insertion or deletion, often referred to as a *butt splice*. If the butt splice occurs at a point in the audio recording that is nearly silent, the butt splice edit may be essentially inaudible, but if the splice occurs during a louder passage of the recording, there may be telltale audible effects and discontinuities. Nevertheless, there may be detectable signal alterations due to the splice that can be observed in the waveform and/or the spectrogram even if there is minimal audible effect.

For example, consider the example recording of Fig. 5.4. The upper panel shows the time waveform, and the lower panel shows the spectrogram. The recording includes some speech utterances and background sounds.

If a forger wanted to remove a particular section of this recording with a butt splice edit, such as the portion of the recording indicated with dashed lines in Fig. 5.5, the result is depicted in Fig. 5.6.

Note that because the edit left an abrupt discontinuity in the waveform, the resulting waveform and spectrogram show evidence of a "click" in the signal. At this time scale, the effect is most easily seen as a vertical line in the spectrogram: the abrupt splice discontinuity in the waveform exposes spectral energy at all frequencies for a brief instant.

Zooming in on the butt splice, it is possible to see the abrupt change in the signal waveform caused by the deletion, as seen in Fig. 5.7. The abrupt change in the time domain waveform leads to the spread of high-frequency energy at the corresponding point in the spectrogram.

While this example makes it appear easy to identify a possible butt splice edit, a more skilled forger would conceal the edit by choosing the edit point in order to minimize the signal discontinuity and by using a short *cross-fade* instead of the butt splice. The cross-fade means overlapping a few samples from before the edit and a few samples after the edit and tapering the amplitude to blend the samples, thereby reducing the likelihood of a noticeable discontinuity at the splice point.

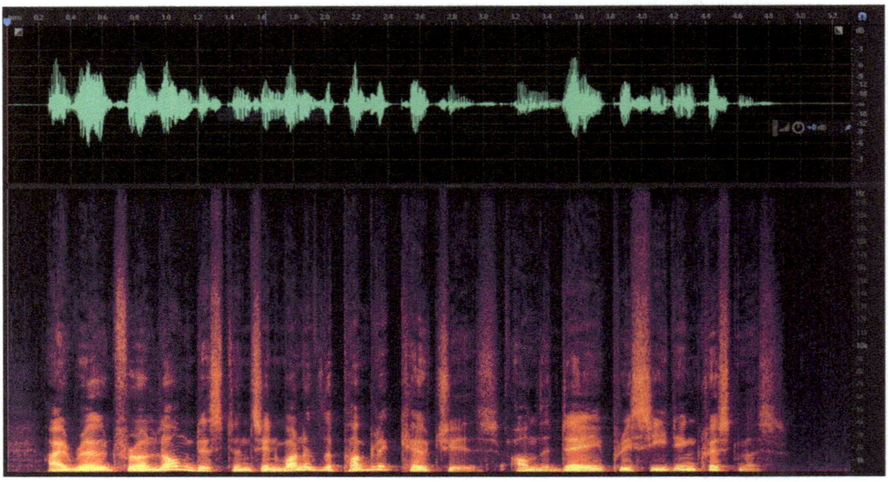

Fig. 5.4 Example audio recording of speech. Upper panel: signal waveform. Lower panel: signal spectrogram (overall duration 5.3 s, frequency range 0–22 kHz, linear scale)

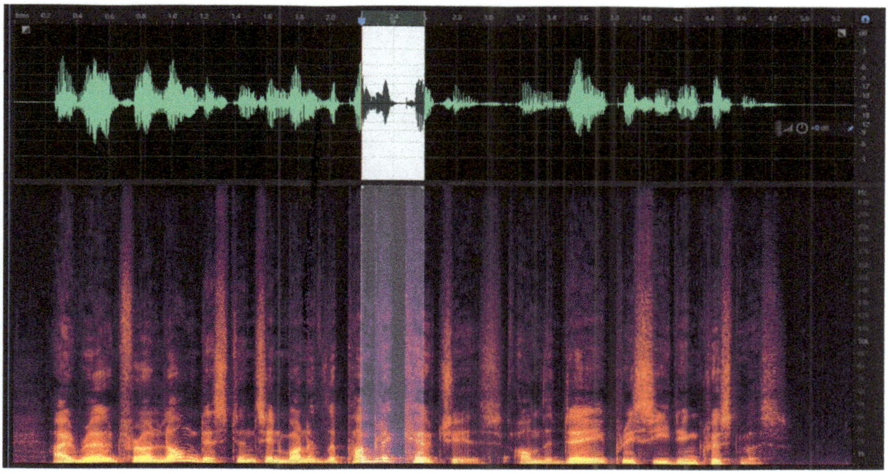

Fig. 5.5 Signal portion (0.4 s) to be removed by butt splice deletion (overall duration 5.3 s, frequency range 0–22 kHz, linear scale)

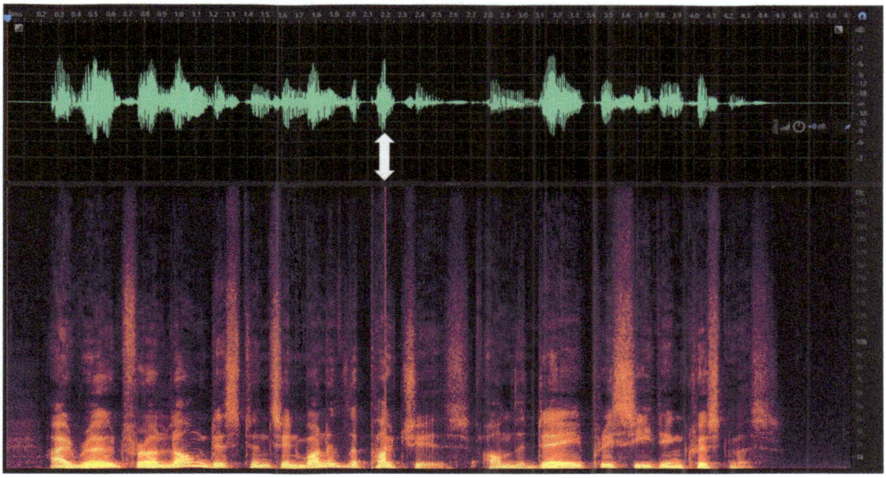

Fig. 5.6 Signal after highlighted portion of Fig. 5.5 is removed (overall duration 4.9 s, frequency range 0–22 kHz, linear scale)

Performing the same deletion edit depicted in Fig. 5.5, but with a 2-ms cross-fade instead of the butt splice, conceals the waveform effect in this example, as shown in Fig. 5.8 and enlarged in Fig. 5.9.

A forger may also attempt to introduce new material into an existing recording, either by opening up a gap in the file for the insertion or by additively mixing the contrived material into the recording. As with the deletion, the boundaries of the insertion could be a butt splice or a concealed cross-fade. If the forger uses skill and care, the edit points may be virtually undetectable in the waveform itself.

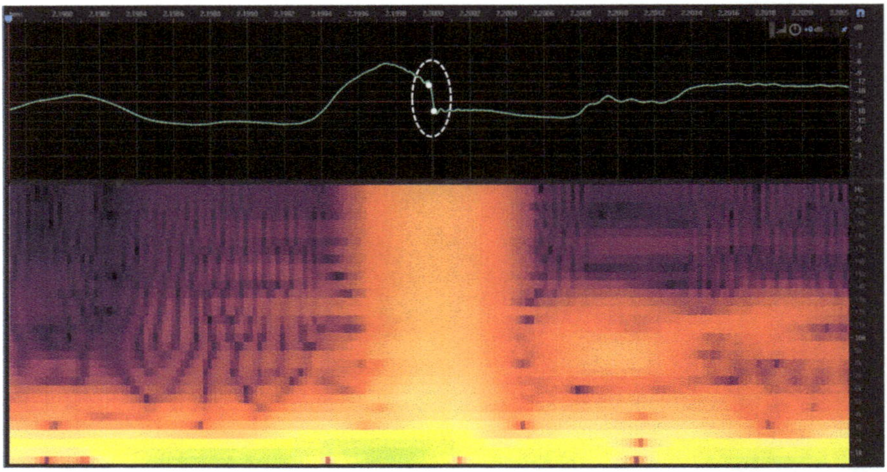

Fig. 5.7 Enlargement of time interval containing the butt splice discontinuity (overall duration 4.8 ms, frequency range 0–22 kHz, linear scale)

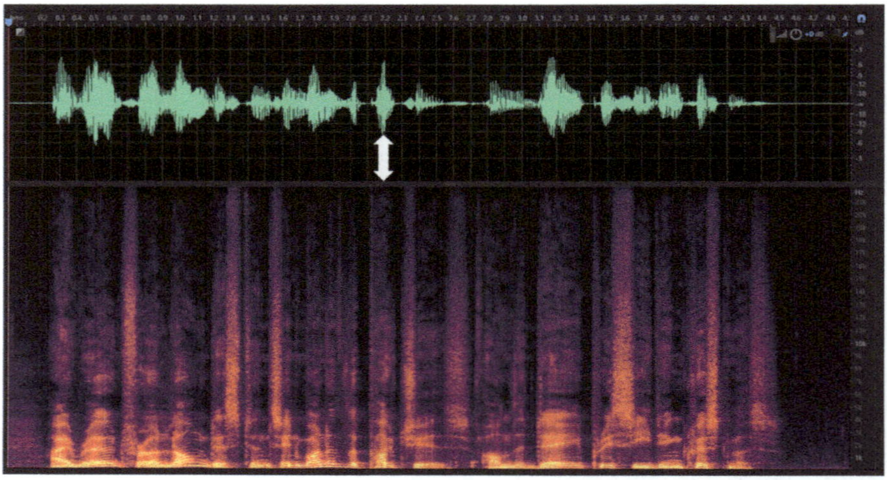

Fig. 5.8 Edit region of Fig. 5.5, but with 2 ms cross-fade instead of the simple butt splice (overall duration 4.9 s, frequency range 0–22 kHz, linear scale)

5.2.2 Other Authenticity Observations

While a smooth edit may reduce the likelihood of first-order detection of an alteration, there may still be signal observations that could raise questions about authenticity. These include background sounds, reverberation, and other acoustic information present in the recording.

In assessing a continuous recording, the audio forensic examiner can observe the acoustic reverberation and background sound level and detect whether there are any

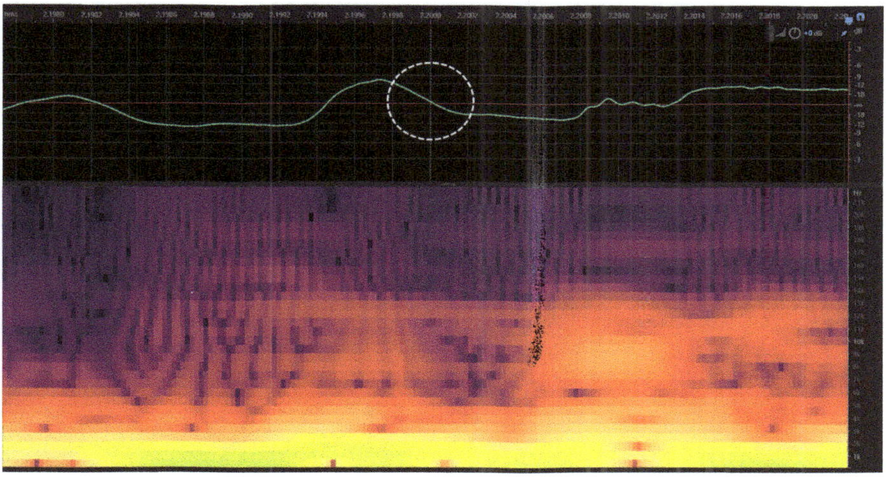

Fig. 5.9 Enlargement of time interval of Fig. 5.8 containing the cross-fade edit (overall duration 4.8 ms, frequency range 0–22 kHz, linear scale)

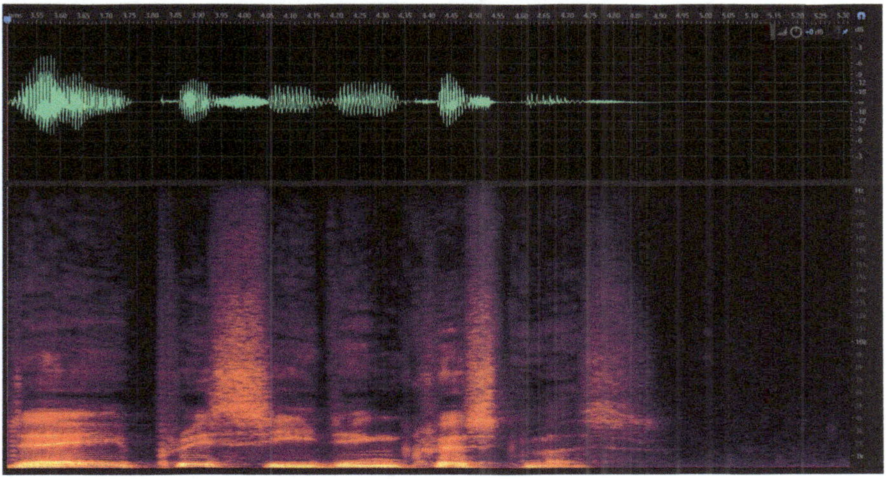

Fig. 5.10 Speech recording with little reverberation (overall duration 1.8 s, frequency range 0–22 kHz, linear scale)

unexplained changes in the background characteristics that could indicate a deletion or an insertion. As noted previously, a recording microphone picks up the direct sound of a source, such as a person talking, but also picks up the acoustic reflections of that sound source from the floor, walls, and other nearby surfaces. The microphone will also pick up any other sounds in the recording environment, such as wind, doors closing, mechanical sounds and alarms, etc.

For example, Fig. 5.10 shows a segment of speech recorded in a room with little reverberation, often referred to as a "dry" recording environment.

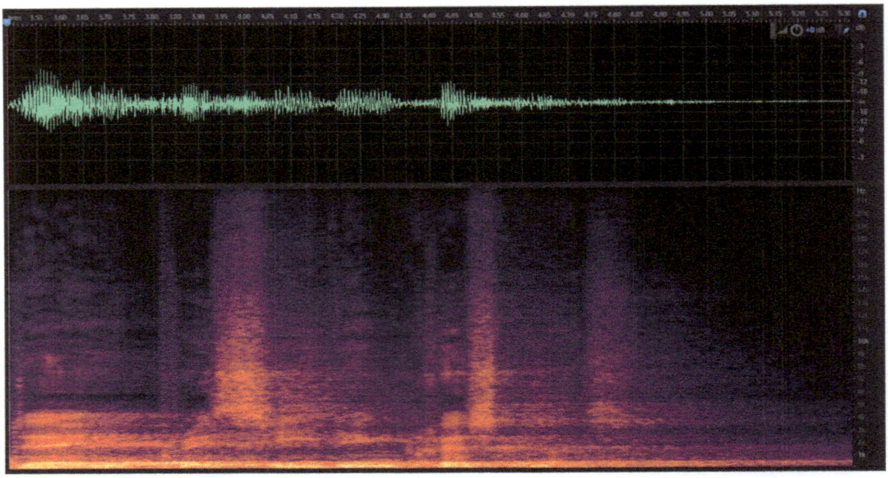

Fig. 5.11 Speech recording with strong reverberation present (overall duration 1.8 s, frequency range 0–22 kHz, linear scale)

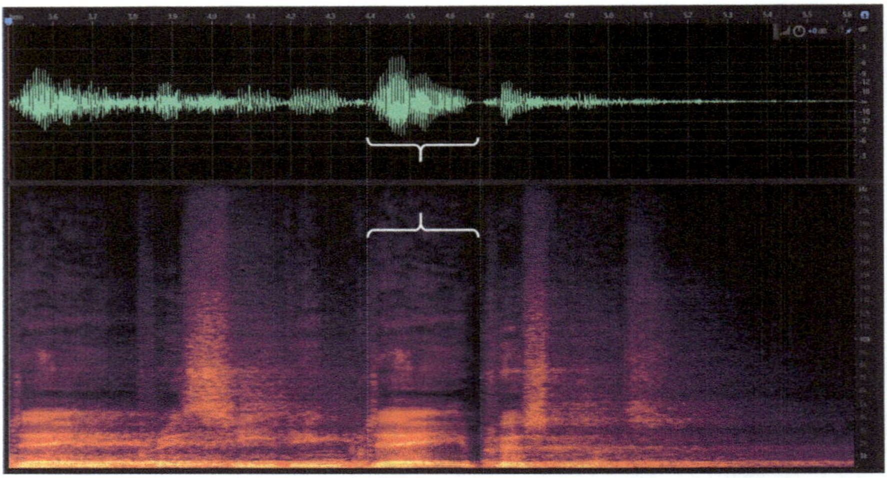

Fig. 5.12 Reverberant recording of Fig. 5.11 with a "dry" insertion (overall duration 2.1 s, frequency range 0–22 kHz, linear scale)

The gaps (dark areas) between the uttered words are particularly visible in the spectrogram, and the lack of background noise and reverberation is apparent. A recording of speech with reverberation present is shown in Fig. 5.11. The gaps between words visible in Fig. 5.10 are now filled with the lingering echoes and reverberation of the preceding sounds.

If there were an attempt to insert newly created material into a reverberant recording such as Fig. 5.11, the forgery could show a change in the reverberation pattern in the spectrogram, as well as in critical listening. Figure 5.12 shows an example in which a short utterance of dry speech is inserted into the recording of

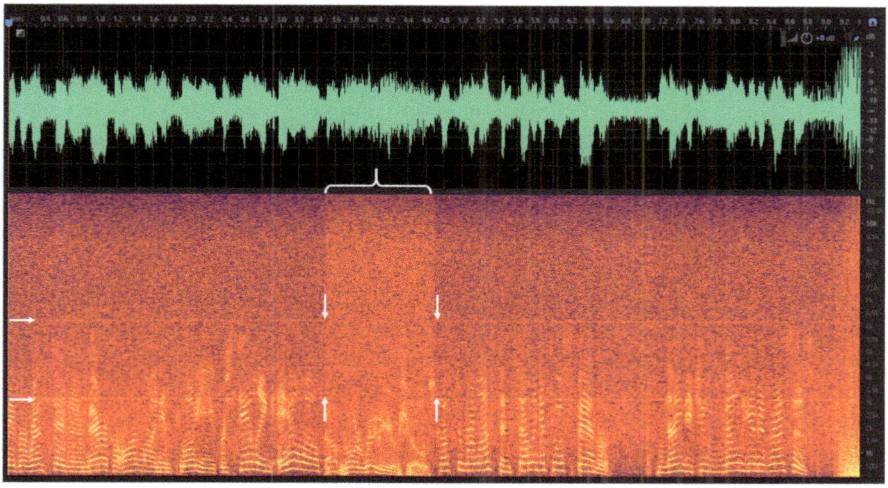

Fig. 5.13 Noisy speech recording with an apparent insertion. Note gap in continuous tones (horizontal lines) denoted by arrows and change in spectral "texture" (overall duration 9.3 s, frequency range 0–11 kHz, linear scale)

Fig. 5.11. The inserted speech lacks the reverb tail apparent after the other recorded words, indicating a likely edit point.

Another example is shown in Fig. 5.13. As seen in the spectrogram, the recording has significant background noise and two continuous discrete tones (horizontal lines indicated by the arrows on the left). In this example, there is a brief section indicated with a subtle difference in noise texture and the absence of the tones. These observations indicate a likely edit insertion into the recording.

5.2.3 Electrical Network Frequency (ENF) Analysis

An interesting potential technique for audio forensic investigations involves a particular background sound that may be present in a recording: the residual "hum" of the electrical power network. This "hum" is usually considered undesirable interference, but there are potentially some possibilities for using this background sound to assess authenticity.

The electrical network frequency (ENF) in the United States and some other countries is nominally 60 Hz, and 50 Hz ENF is common in Europe and many other parts of the world. The operation of the contemporary electrical power system requires that all of the AC electrical generators interconnected cooperatively through the electrical power network, or "the grid," operate synchronously: all of the 60 Hz power waveforms anywhere in the electrical grid are kept at exactly the same frequency and in-phase with each other. The United States power network is comprised of three large grids: eastern grid, western grid, and Texas. Within each grid, the power frequency is the same at every generator and outlet.

The electrical grid operating organization has to control the power system so that the amount of electricity being generated exactly matches the amount of electricity needed at any point in time, which keeps the ENF at the 60 Hz nominal value. However, if electrical use declines at a given time, the rotating electrical generators have less load and tend to turn a bit faster, increasing the ENF. On the other hand, if the demand for electricity increases, the electrical generators have a greater load and tend to slow down, decreasing the ENF. The grid operating organization must keep the variation to within about ±0.5 Hz by generating more or less power as needed, and the precise ENF frequency fluctuates gradually and unpredictably within the allowable range.

Because all of the generators attached to the grid operate synchronously, the instantaneous ENF will be the same everywhere on the entire electrical grid. If an audio recording includes hum from the electrical power system, the frequency of the hum is the electrical network frequency, and therefore it should be possible—at least in principle—to compare the recorded ENF fluctuations with a database of known power grid ENF measurements to identify the date and time of the recording.

Audio recording systems are generally designed to minimize the effects of AC (alternating current) power line interference, but low levels of residual power signals may appear in the audio circuitry and become part of the audio recording. This is most likely to occur when a line transformer powers the recording device, but some residual line frequency pickup is possible even with battery-powered equipment if the recording device is susceptible to the magnetic fields emanating from nearby wiring (Brixen 2007, 2008; Grigoras 2005, 2007).

In addition to needing a reference power grid frequency database, ENF analysis requires several important assumptions and measurements.

First, the recording must contain a detectable hum signal of sufficient strength that its precise frequency can be determined several times per second. The extraction process can be difficult because the 60 Hz ENF (and its harmonics) is within the regular audio bandwidth, so there may also be acoustic signals in the same frequency range as the ENF.

Second, the length of the recording and the corresponding duration of the ENF record need to be sufficiently long that the extracted ENF pattern is reliably distinguishable from any other span of time.

Third, the extracted ENF depends upon the actual sampling rate (or analog recording speed) of the audio signal, and any discrepancy in the recording process will introduce a systematic frequency shift.

An example procedure for extracting the ENF from audio recordings is shown in Fig. 5.14 (Cooper 2008). An example comparison of ENF data obtained from an audio recording and the reference ENF data from the electrical power system are shown in Fig. 5.15.

5.2.4 Metadata Consistency

Contemporary digital audio recordings are stored as computer files in a number of standard or proprietary formats. The audio file format includes the bytes containing the digital audio data, along with additional useful information *about* the recording, known

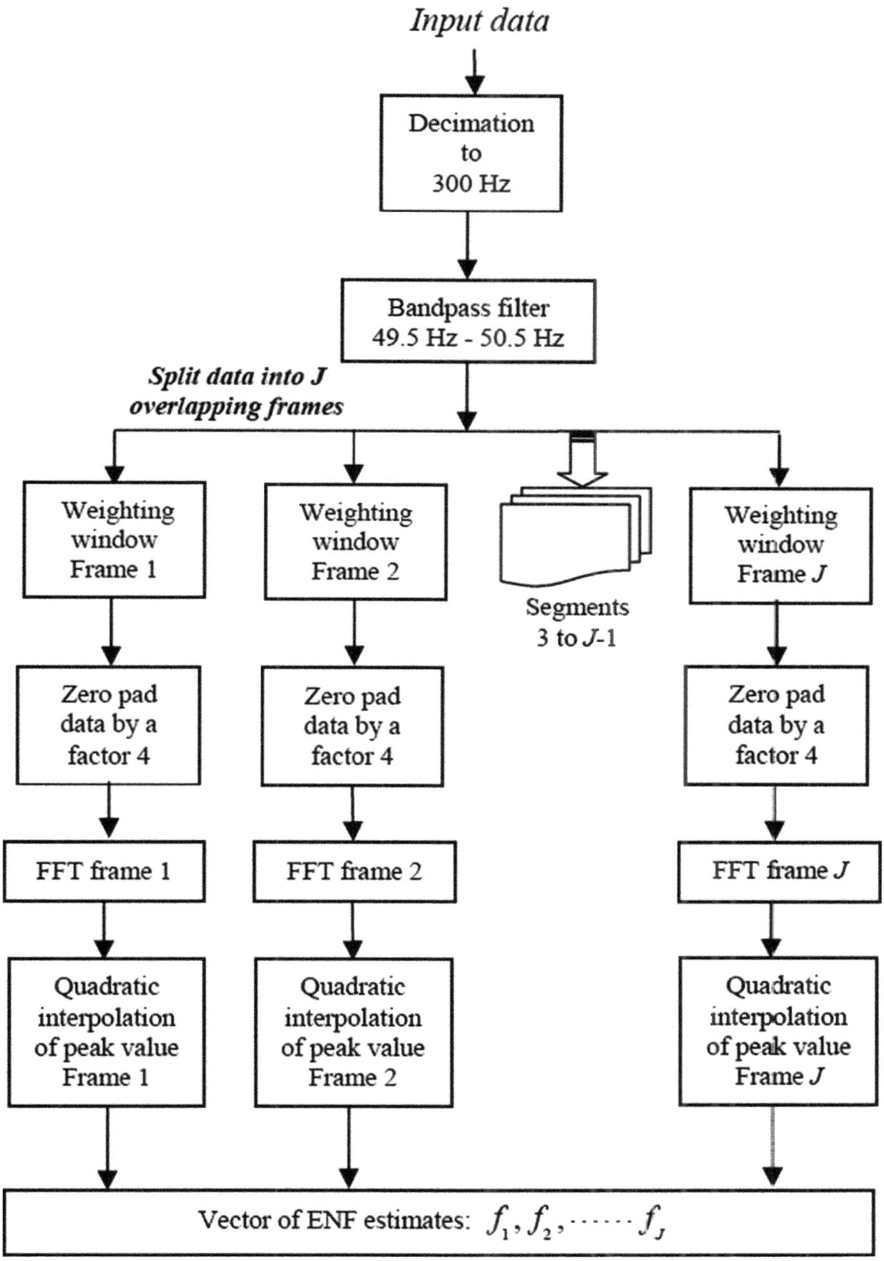

Fig. 5.14 Proposed ENF processing procedure (from Cooper 2008, reprinted by permission)

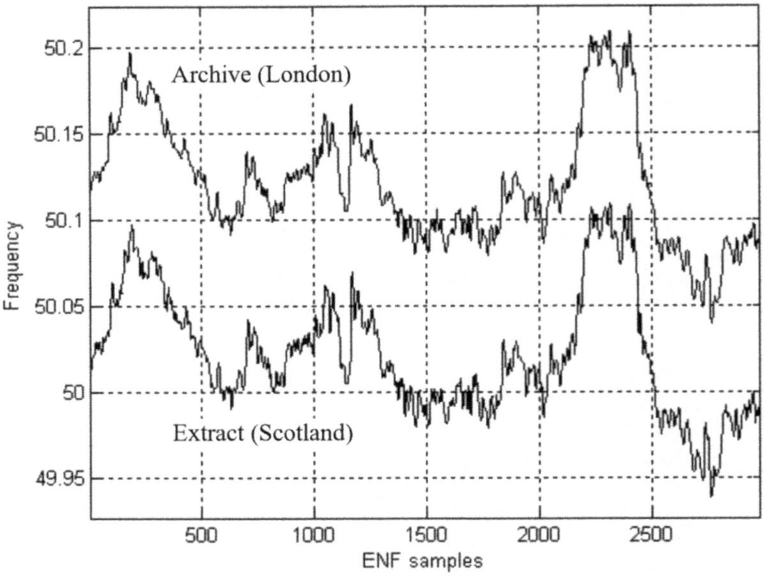

Fig. 5.15 Automated match between extracted ENF and database. The total time represented is approximately 70 min (1.4 s per ENF sample). The extracted waveform has been offset by 0.1 Hz to aid visual comparison (from Cooper 2008, reprinted by permission)

as *metadata*. Metadata included in an audio file format might include the sampling rate, the number of audio channels, the brand and model number of the recording device, the date of the recording, and so forth (Koenig and Lacey 2014). A forensic examiner can nondestructively observe the metadata using an editor program capable of displaying the binary information as readable characters (typically hexadecimal notation).

An example listing of the file metadata for an example MP3 audio file is shown in Fig. 5.16. The information is from a binary display program and shows the file contents as text characters in the right column, and the binary data values are given as hexadecimal (base-16) values in the body of the figure. Note that in the right column, there are recognizable strings of characters such as "ID3," identifying the beginning of the file to be containing standard "tags" from id3.org. The file was recorded by an Olympus brand model 702 memo recorder on July 11, 2018. Note the strings OLY (Olympus), mp3 (recording mode), 702 (recorder model), and 20,180,711 (date code). Other bytes in the metadata header may also be meaningful if the manufacturer provides technical information regarding the recording device. Often this information is proprietary, and its meaning can only be determined empirically.

When a recording device opens a file, performs the recording operation, and then saves and closes the file, the device also updates the metadata. An authentic recording will have metadata that is consistent with the stated circumstances of the recording and the file contents. As in the example above, the metadata must match the type and model of recorder, the recording date, and the recording duration.

```
          00 01 02 03  04 05 06 07  08 09 0A 0B  0C 0D 0E 0F  0123456789ABCDEF
00 0000:  49 44 33 03  00 00 00 00  0B 76 58 4F  4C 59 00 00  ID3......vXOLY..
00 0010:  05 EC 00 00  03 6D 70 33  02 00 01 00  02 00 00 00  .....mp3........
00 0020:  37 30 32 5F  20 20 20 20  20 20 20 20  20 20 20 20  702_
00 0030:  0F 00 00 00  FE FF FF FF  FF FF 31 38  30 37 31 31  ..........180711
00 0040:  31 36 30 30  34 39 31 38  30 37 31 31  31 36 31 30  1600491807111610
00 0050:  35 34 30 30  31 30 30 34  FF 07 FF FF  FF FF FF FF  54001004........
00 0060:  FF FF FF FF  FF FF FF FF  FF FF 3C 38  09 00 43 FF  ..........<8..C.
00 0070:  FF FF FF FF  FF FF FF FF  FF FF FF FF  FF FF 00 00  ................
00 0080:  00 00 00 00  00 00 00 00  00 00 00 00  00 00 00 00  ................
00 0090:  00 00 00 00  00 00 00 00  00 00 00 00  00 00 00 00  ................
00 00A0:  00 00 00 00  00 00 00 00  00 00 00 00  00 00 00 00  ................
00 00B0:  00 00 00 00  00 00 00 00  00 00 00 00  00 00 00 00  ................
00 00C0:  00 00 00 00  00 00 00 00  00 00 00 00  00 00 00 00  ................
00 00D0:  00 00 00 00  00 00 00 00  00 00 00 00  00 00 00 00  ................
00 00E0:  00 00 FF 00  00 00 00 00  00 00 00 00  00 00 00 00  ................
00 00F0:  00 00 00 00  00 00 00 00  00 00 00 00  00 00 00 00  ................
00 0100:  00 00 00 00  00 00 00 00  00 00 00 00  00 00 00 00  ................
00 0110:  00 00 00 00  00 00 00 00  00 00 00 00  00 00 00 00  ................
00 0120:  00 00 00 00  00 00 00 00  00 00 00 00  00 00 00 00  ................
00 0130:  00 00 00 00  00 00 00 00  00 00 00 00  00 00 00 00  ................
00 0140:  00 00 00 00  00 00 00 00  00 00 00 00  63 00 00 00  ............c...
00 0150:  FF FF FF FF  FF FF FF FF  FF FF FF FF  FF FF FF FF  ................
00 0160:  FF FF FF FF  FF FF FF FF  FF FF FF FF  FF FF FF FF  ................
00 0170:  FF FF FF FF  FF FF FF FF  FF FF FF FF  FF FF FF FF  ................
00 0180:  FF FF FF FF  FF FF FF FF  FF FF FF FF  FF FF FF FF  ................
00 0190:  FF FF FF FF  FF FF FF FF  FF FF FF FF  FF FF FF FF  ................
00 01A0:  FF FF FF FF  FF FF FF FF  FF FF FF FF  FF FF FF FF  ................
00 01B0:  FF FF FF FF  FF FF FF FF  FF FF FF FF  FF FF FF FF  ................
00 01C0:  FF FF FF FF  FF FF FF FF  FF FF FF FF  FF FF FF FF  ................
00 01D0:  FF FF FF FF  FF FF FF FF  FF FF FF FF  FF FF FF FF  ................
00 01E0:  FF FF FF FF  FF FF FF FF  FF FF FF FF  FF FF FF FF  ................
00 01F0:  FF FF FF FF  FF FF FF FF  FF FF FF FF  FF FF FF FF  ................
00 0200:  FF FF FF FF  FF FF FF FF  FF FF FF FF  FF FF FF FF  ................
00 0210:  FF FF FF FF  FF FF FF FF  FF FF FF FF  FF FF FF FF  ................
00 0220:  FF FF FF FF  FF FF FF FF  FF FF FF FF  FF FF FF FF  ................
00 0230:  FF FF FF FF  FF FF FF FF  FF FF FF FF  FF FF FF FF  ................
00 0240:  FF FF FF FF  FF FF FF FF  FF FF FF FF  FF FF FF FF  ................
00 0250:  FF FF FF FF  FF FF FF FF  FF FF FF FF  FF FF FF FF  ................
00 0260:  FF FF FF FF  FF FF FF FF  FF FF FF FF  FF FF FF FF  ................
00 0270:  FF FF FF FF  FF FF FF FF  FF FF FF FF  FF FF FF FF  ................
00 0280:  FF FF FF FF  FF FF FF FF  FF FF FF FF  FF FF FF FF  ................
00 0290:  FF FF FF FF  FF FF FF FF  FF FF FF FF  FF FF FF FF  ................
00 02A0:  FF FF FF FF  FF FF FF FF  FF FF FF FF  FF FF FF FF  ................
00 02B0:  FF FF FF FF  FF FF FF FF  FF FF FF FF  FF FF FF FF  ................
00 02C0:  FF FF FF FF  FF FF FF FF  FF FF FF FF  FF FF FF FF  ................
00 02D0:  FF FF FF FF  FF FF FF FF  FF FF FF FF  FF FF FF FF  ................
00 02E0:  FF FF FF FF  FF FF FF FF  FF FF FF FF  FF FF FF FF  ................
00 02F0:  FF FF FF FF  FF FF FF FF  FF FF FF FF  FF FF FF FF  ................
00 0300:  FF FF FF FF  FF FF FF FF  FF FF FF FF  FF FF FF FF  ................
00 0310:  FF FF FF FF  FF FF FF FF  FF FF FF FF  FF FF FF FF  ................
00 0320:  FF FF FF FF  FF FF FF FF  FF FF FF FF  FF FF FF FF  ................
00 0330:  FF FF FF FF  FF FF FF FF  FF FF FF FF  FF FF FF FF  ................
00 0340:  FF FF FF FF  FF FF FF FF  FF FF FF FF  FF FF FF FF  ................
00 0350:  FF FF FF FF  FF FF FF FF  FF FF FF FF  FF FF FF FF  ................
```

Fig. 5.16 Example of metadata in a digital audio file, displayed as hexadecimal and text. This file is from an Olympus brand model 702 memo recorder, and the file was created on July 11, 2018. Note the strings "ID3," "OLY," "mp3," "702," and the date code "20,180,711"

However, if an authentic original file is transferred to another device or to a computer, edited on that device, and then saved to a new file, the editing device or software package typically updates various details in the metadata. For example, the file depicted in Fig. 5.16 was subsequently opened with the *Adobe Audition*® software package and then saved with a new file name, resulting in the metadata shown in Fig. 5.17. The software has clearly altered the metadata at the head of the file.

```
           00 01 02 03  04 05 06 07  08 09 0A 0B  0C 0D 0E 0F  0123456789ABCDEF
00 0000:   49 44 33 03  00 00 00 00  30 01 54 59  45 52 00 00  ID3.....0.TYER..
00 0010:   00 06 00 00  00 32 30 31  38 00 54 44  41 54 00 00  ......2018.TDAT..
00 0020:   00 06 00 00  00 31 32 30  37 00 54 49  4D 45 00 00  ......1207.TIME..
00 0030:   00 06 00 00  00 31 36 34  39 00 50 52  49 56 00 00  ......1649.PRIV..
00 0040:   0F C7 00 00  58 4D 50 00  3C 3F 78 70  61 63 6B 65  ....XMP.<?xpacke
00 0050:   74 20 62 65  67 69 6E 3D  22 EF BB BF  22 20 69 64  t begin="..." id
00 0060:   3D 22 57 35  4D 30 4D 70  43 65 68 69  48 7A 72 65  ="W5M0MpCehiHzre
00 0070:   53 7A 4E 54  63 7A 6B 63  39 64 22 3F  3E 0A 3C 78  SzNTczkc9d"?>.<x
00 0080:   3A 78 6D 70  6D 65 74 61  20 78 6D 6C  6E 73 3A 78  :xmpmeta xmlns:x
00 0090:   3D 22 61 64  6F 62 65 3A  6E 73 3A 6D  65 74 61 2F  ="adobe:ns:meta/
00 00A0:   22 20 78 3A  78 6D 70 74  6B 3D 22 41  64 6F 62 65  " x:xmptk="Adobe
00 00B0:   20 58 4D 50  20 43 6F 72  65 20 35 2E  36 2D 63 31   XMP Core 5.6-c1
00 00C0:   34 34 20 37  39 2E 31 36  32 30 34 35  2C 20 32 30  44 79.162045, 20
00 00D0:   31 38 2F 30  31 2F 32 33  2D 30 36 3A  30 35 3A 35  18/01/23-06:05:5
00 00E0:   32 20 20 20  20 20 20 20  20 22 3E 0A  20 3C 72 64  2        ">. <rd
00 00F0:   66 3A 52 44  46 20 78 6D  6C 6E 73 3A  72 64 66 3D  f:RDF xmlns:rdf=
00 0100:   22 68 74 74  70 3A 2F 2F  77 77 77 2E  77 33 2E 6F  "http://www.w3.o
00 0110:   72 67 2F 31  39 39 39 2F  30 32 2F 32  32 2D 72 64  rg/1999/02/22-rd
00 0120:   66 2D 73 79  6E 74 61 78  2D 6E 73 23  22 3E 0A 20  f-syntax-ns#">.
00 0130:   20 3C 72 64  66 3A 44 65  73 63 72 69  70 74 69 6F   <rdf:Descriptio
00 0140:   6E 20 72 64  66 3A 61 62  6F 75 74 3D  22 22 0A 20  n rdf:about="".
00 0150:   20 20 20 78  6D 6C 6E 73  3A 78 6D 70  44 4D 3D 22     xmlns:xmpDM="
00 0160:   68 74 74 70  3A 2F 2F 6E  73 2E 61 64  6F 62 65 2E  http://ns.adobe.
00 0170:   63 6F 6D 2F  78 6D 70 2F  31 2E 30 2F  44 79 6E 61  com/xmp/1.0/Dyna
00 0180:   6D 69 63 4D  65 64 69 61  2F 22 0A 20  20 20 20 78  micMedia/".    x
00 0190:   6D 6C 6E 73  3A 78 6D 70  3D 22 68 74  74 70 3A 2F  mlns:xmp="http:/
00 01A0:   2F 6E 73 2E  61 64 6F 62  65 2E 63 6F  6D 2F 78 61  /ns.adobe.com/xa
00 01B0:   70 2F 31 2E  30 2F 22 0A  20 20 20 20  78 6D 6C 6E  p/1.0/".    xmln
00 01C0:   73 3A 78 6D  70 4D 4D 3D  22 68 74 74  70 3A 2F 2F  s:xmpMM="http://
00 01D0:   6E 73 2E 61  64 6F 62 65  2E 63 6F 6D  2F 78 61 70  ns.adobe.com/xap
00 01E0:   2F 31 2E 30  2F 6D 6D 2F  22 0A 20 20  20 20 78 6D  /1.0/mm/".    xm
00 01F0:   6C 6E 73 3A  73 74 45 76  74 3D 22 68  74 74 70 3A  lns:stEvt="http:
00 0200:   2F 2F 6E 73  2E 61 64 6F  62 65 2E 63  6F 6D 2F 78  //ns.adobe.com/x
00 0210:   61 70 2F 31  2E 30 2F 73  54 79 70 65  2F 52 65 73  ap/1.0/sType/Res
00 0220:   6F 75 72 63  65 45 76 65  6E 74 23 22  0A 20 20 20  ourceEvent#".
00 0230:   20 78 6D 6C  6E 73 3A 64  63 3D 22 68  74 74 70 3A   xmlns:dc="http:
00 0240:   2F 2F 70 75  72 6C 2E 6F  72 67 2F 64  63 2F 65 6C  //purl.org/dc/el
00 0250:   65 6D 65 6E  74 73 2F 31  2E 31 2F 22  0A 20 20 20  ements/1.1/".
00 0260:   78 6D 70 3A  4D 65 74 61  64 61 74 61  44 61 74 65  xmp:MetadataDate
00 0270:   3D 22 32 30  31 38 2D 30  37 2D 31 32  54 31 36 3A  ="2018-07-12T16:
00 0280:   34 39 3A 33  33 2D 30 36  3A 30 30 22  0A 20 20 20  49:33-06:00".
00 0290:   78 6D 70 3A  43 72 65 61  74 6F 72 54  6F 6F 6C 3D  xmp:CreatorTool=
00 02A0:   22 41 64 6F  62 65 20 41  75 64 69 74  69 6F 6E 20  "Adobe Audition
00 02B0:   43 43 20 32  30 31 38 2E  31 20 28 57  69 6E 64 6F  CC 2018.1 (Windo
00 02C0:   77 73 29 22  0A 20 20 20  78 6D 70 3A  43 72 65 61  ws)".   xmp:Crea
00 02D0:   74 65 44 61  74 65 3D 22  32 30 31 38  2D 30 37 2D  teDate="2018-07-
00 02E0:   31 32 54 31  36 3A 34 39  3A 33 33 2D  30 36 3A 30  12T16:49:33-06:0
00 02F0:   30 22 0A 20  20 20 78 6D  70 3A 4D 6F  64 69 66 79  0".   xmp:Modify
00 0300:   44 61 74 65  3D 22 32 30  31 38 2D 30  37 2D 31 32  Date="2018-07-12
00 0310:   54 31 36 3A  34 39 3A 33  33 2D 30 36  3A 30 30 22  T16:49:33-06:00"
00 0320:   0A 20 20 20  78 6D 70 4D  4D 3A 49 6E  73 74 61 6E  .   xmpMM:Instan
00 0330:   63 65 49 44  3D 22 78 6D  70 2E 69 69  64 3A 35 61  ceID="xmp.iid:5a
00 0340:   31 62 39 30  32 36 2D 62  65 31 37 2D  32 39 34 66  1b9026-be17-294f
00 0350:   2D 39 31 61  36 2D 37 61  39 34 64 36  34 30 38 38  -91a6-7a94d64088
```


Fig. 5.17 Example of metadata for the file of Fig. 5.16 opened and saved with a different software package, thereby altering the metadata

If the examination reveals inconsistency between the metadata and the expected circumstances of the recording, this could indicate that the recording was edited or modified in some manner, possibly representing a forgery.

As explained previously, one of the difficulties associated with digital files is the inability to distinguish between an authentic file and a forgery prepared by a skillful adversary. This caveat applies to metadata as well, since a forgery could have metadata altered in such a manner as to appear consistent with an authentic recording. Thus, an examiner is generally only able to report upon inconsistencies that could indicate inauthenticity, not to guarantee authenticity if no inconsistencies are found.

References

Advisory Panel on White House Tapes. (1974). *The executive office building tape of June 20, 1972: Report on a technical investigation.* Washington, D.C.: United States District Court for the District of Columbia.

Audio Engineering Society. (2000). *AES43-2000: AES standard for forensic purposes – Criteria for the authentication of analog audio tape recordings*. New York: AES.

Brixen, E. B. (2007). Techniques for the authentication of digital audio recordings. In *Proceedings Audio Engineering Society 122nd Convention, Vienna, Austria, Convention paper 7014.*

Brixen, E. B. (2008). ENF–Quantification of the magnetic field. In *Proceedings Audio Engineering Society 33rd Conference, Audio Forensics—Theory and Practice, Denver, CO* (pp. 1–6).

Begault, D. R., Brustad, B. M., & Stanley, A. M. (2005) Tape analysis and authentication using multi-track recorders. In *Proceedings Audio Engineering Society 26th Conference, Audio Forensics in the Digital Age, Denver, CO* (pp. 115–121).

Cooper, A. J. (2008). The electric network frequency (ENF) as an aid to authenticating forensic digital audio recordings – An automated approach. In *Proceedings Audio Engineering Society 33rd Conference, Audio Forensics—Theory and Practice, Denver, CO* (pp. 1–10).

Grigoras, C. (2005). Digital audio recording analysis: The electric network frequency (ENF) criterion. *International Journal of Speech, Language and the Law, 12*(1), 63–76.

Grigoras, C. (2007). Application of ENF analysis method in authentication of digital audio and video recordings, In *Proceedings Audio Engineering Society 123rd Convention, New York, NY, Convention paper 1273.*

Koenig, B. E. (1990). Authentication of forensic audio recordings. *Journal of the Audio Engineering Society, 38*(1/2), 3–33.

Koenig, B. E., & Lacey, D. S. (2014). Forensic authenticity analyses of the metadata in re-encoded WAV files. In *Proceedings of the Audio Engineering Society 54th International Conference: Audio Forensics, London, U.K.*

Marr, K., & Pappas, D. P. (2008). Magneto-resistive field mapping of analog audio tapes for forensics imaging. In *Proceedings of the Audio Engineering Society 33rd Conference, Audio Forensics Theory and Practice, Denver, CO* (pp. 1–7).

Chapter 6
Audio Signal Enhancement

Popular television shows and movies involving law enforcement and forensic analysis sometimes include an example of a terribly scratchy and obliterated audio recording that is magically transformed into pristine sound, leading to the guilty culprit. Other plots might have a blurry handheld snapshot from which the technician pulls out a perfect image of a license plate. These fictional examples of forensic enhancement make for great entertainment. Noisy forensic audio evidence in the real world seldom affords a basis for magical perfection, but there are important methods for audio enhancement that can provide very useful improvements for forensic purposes.

Among the most common audio enhancement tasks for forensic audio examiners is a request to improve the intelligibility of a conversation from a poor-quality surveillance recording obtained with a hidden microphone. Forensic audio recordings often occur in non-ideal circumstances, and therefore the recordings often contain noise, clipping, distortion, interfering sounds, and other shortcomings that can affect the quality and intelligibility of speech and impede analysis of background sounds and other subtleties. The surreptitious aspect of the recording process often prevents good microphone placement, leading to highly reverberant recordings and interfering noise such as the microphone rubbing against clothing.

Forensic audio enhancement is generally performed off-line using a certified digital copy of the evidentiary recording. The original evidence is not altered. The enhancement is accomplished iteratively so that the examiner can listen to the results and make adjustments systematically and gradually.

6.1 Enhancement Assessment

The audio forensic examiner develops a preliminary impression about the audio quality and speech intelligibility during the *initial aural evaluation* and *critical listening* phases. The preliminary impression is important when receiving the client's request, so that everyone will have consistent expectations and goals.

R. C. Maher, *Principles of Forensic Audio Analysis*, Modern Acoustics and Signal Processing, https://doi.org/10.1007/978-3-319-99453-6_6

Among the key observations is the general character of noise and interference in the recording. In some cases, the interfering sound has a consistent character, such as a continuous whine, hum, rumble, or hiss. In this case, the interfering sound is referred to as *stationary noise*. If the stationary noise occupies a frequency range that differs from the signals of interest, such as a speech recording with steady rumble in the frequency range below 100 Hz, it may be possible to apply a fixed filter. In this case, a bandpass filter can be used to pass approximately the expected speech bandwidth 250 Hz to 4 kHz while attenuating the low-frequency noise. If the stationary noise occupies the same frequency range as the desired signal, a simple separation filter will not be feasible, but it may still be possible to apply equalization to improve the audibility/intelligibility of the desired signal.

In other cases, the noise and interference may be time-variant impulsive clicks, rattles, or microphone-related sounds such as wind turbulence or fabric rubbing on the microphone. These *non-stationary noise* sources generally require more complicated processing than stationary noise sources and are often not effectively suppressed.

In some audio forensic investigations, the information of interest may not be the relatively loud *foreground* sounds, but instead the quiet and subtle *background* sounds present during the recording. For example, the investigation may involve a question about the sequence of events preceding a particular utterance, such as footsteps, a door creaking, or the soft sound of a distant voice. Rather than attempting to suppress the background "noise," the effort will be to boost the background sounds while suppressing the high-level foreground sound.

A common issue is a forensic recording containing speech and stationary noise. The request may be to improve the intelligibility for presentation in court or for a stenographer to prepare a written transcript. If the noise and speech signal occupy the same bandwidth, a human listener may actually become accustomed to the stationary noise and be able to discern aspects of the recording that are technically below the *noise floor*. In this situation, adjusting the playback level up and down slightly may help determine a setting that provides the best speech intelligibility.

6.2 Speech: Quality Vs. Intelligibility

Communications systems engineers have studied speech intelligibility since the invention of the telephone in the 1870s. A multidimensional problem, intelligibility depends upon signal level, bandwidth, and signal-to-noise ratio, among other factors.

As a basic example, consider a set of one-syllable rhyming words:

mat, hat, sat, fat, bat, cat, pat, rat, and vat.

The words mean completely different things, but differ primarily in the brief initial consonant sound. See the spectrographic view of a wideband recording of these words in Fig. 6.1.

If the words are recorded in the presence of background noise, it may be difficult for a listener to distinguish between "cat" and "pat" or between "mat" and "vat." Human listeners can often use context and semantics to make a good guess about the noisy word: "The <*blank*> chased the mouse" is more likely to be fulfilled by "cat" than "bat," while the sentence "The child's colorful <*blank*> made her easy to identify in the crowd" is more likely to be "hat" rather than "bat." This is among the reasons that intelligibility is better for sentences than for isolated words (refer back to Chap. 2, Fig. 2.18).

When noise is present, the signal of Fig. 6.1 becomes less distinct, as shown in Fig. 6.2.

If the recording contains noise and is also limited to the typical telephone voice bandwidth (400 Hz–3.4 kHz), the signal representation may be even less distinct. Figure 6.3 shows an example with noise and limited bandwidth.

An important consideration to keep in mind is that listeners typically assess *quality* differently than *intelligibility*: a recording judged to be of good quality may sometimes have lower intelligibility than a recording judged to be of lesser quality (see Sect. 2.5.4). This seemingly paradoxical result is because in some cases a noisy, crackly recording retains speech features such as fricatives and subtle voiced sounds that are lost if the recording is filtered or otherwise smoothed. The listener may prefer the filtered recording from the standpoint of perceived quality, but a test of understanding speech might reveal better performance with the noisy recording.

Therefore, a forensic examiner or an individual who is attempting to transcribe a noisy speech recording may need to try several different signal processing strategies to get the greatest intelligibility.

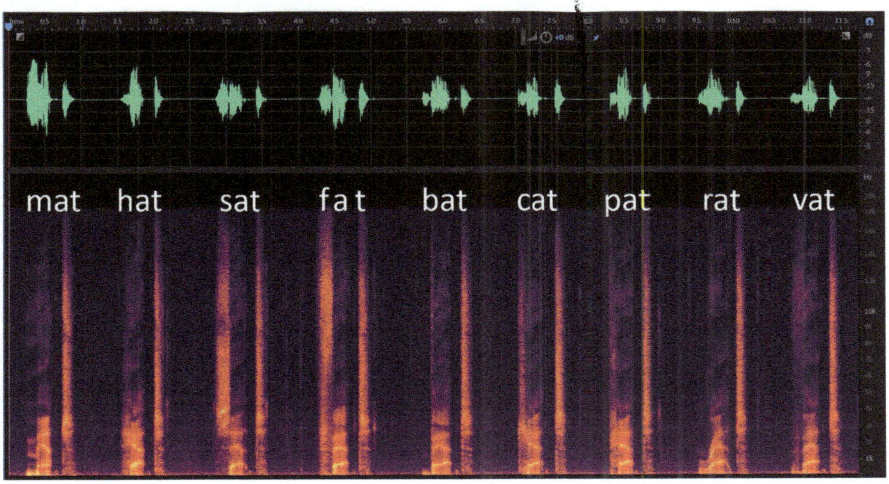

Fig. 6.1 Waveform and spectrogram of a sequence of rhyming words, 48 kHz sampling rate (overall duration 11.5 s, frequency range 0–22 kHz, linear scale)

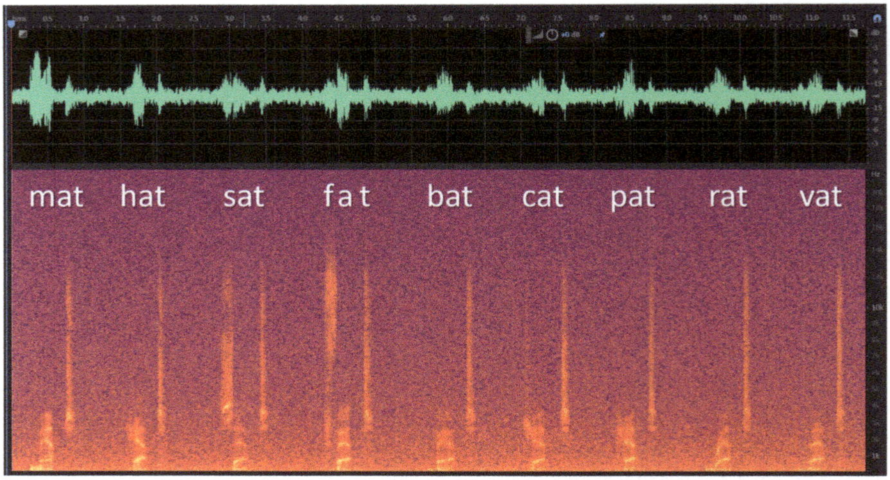

Fig. 6.2 Waveform and spectrum of rhyming words with additive noise, 48 kHz sampling rate (overall duration 11.5 s, frequency range 0–22 kHz, linear scale)

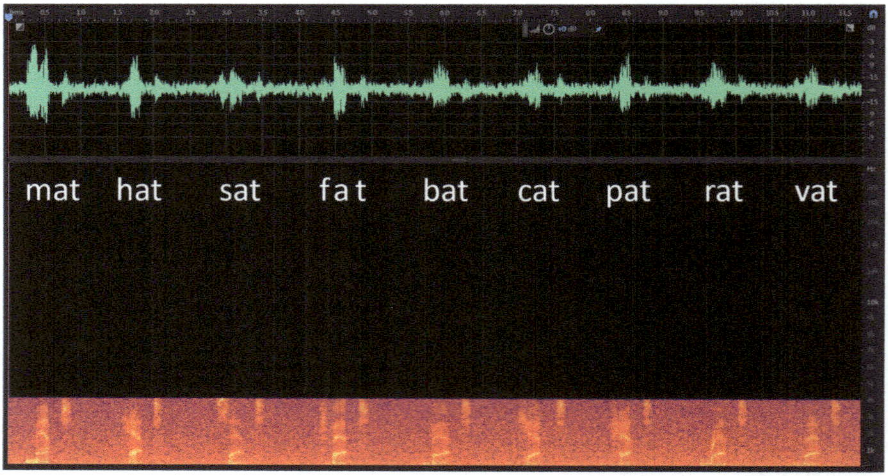

Fig. 6.3 Waveform and spectrogram of rhyming words with noise and bandlimiting: 8 kHz sampling rate, 4 kHz bandwidth (overall duration 11.5 s, frequency range 0–22 kHz, linear scale)

6.3 Techniques for Forensic Audio Enhancement

Unfortunately, there are no perfect methods for improving the quality and/or intelligibility of a noisy forensic recording. Nonetheless, many audio forensic investigations require an examiner to use filtering, gain compression and expansion, click and gap removal, and other techniques to address audible shortcomings in the material, especially if the recording will be presented to novice listeners such as a jury.

6.3.1 Filtering and Equalization

If sections of the recording contain rumble, hum, or audible tones that do not overlap the frequency range of the desired speech or other relevant signals, it is often helpful to try applying an appropriate frequency filter to reduce the out-of-band noise. An audio waveform editor software tool that provides filter capability makes this possible.

Filtering refers to signal processing that selectively emphasizes or de-emphasizes certain frequency ranges in an audio recording. Filtering can be accomplished either with analog circuitry or with digital computation. Filtering is a *linear* operation, meaning that it performs its action passively without needing any prior knowledge about the input signal characteristics. A filter is specified by several parameters, including its bandwidth, its selectivity, and its gain or spectral "shape."

Figure 6.4 depicts a filter that *attenuates* low frequencies and *passes* high frequencies. This is referred to as a *highpass* filter because it allows high frequencies to pass through. A *lowpass* filter, a *bandpass* filter, and a *bandstop* or *notch* filter are shown in Fig. 6.5. The frequency ranges that are attenuated are referred to as the *stopband*, while the frequency ranges that are passed through the filter are called the *passband*.

The *selectivity* of a filter refers to how abruptly the filter's gain changes as a function of frequency in the range between the passband and the stopband.

Equalization also refers to filtering, but the term usually implies that the filter's gain varies in a deliberate manner across the desired passband. Equalization might be more familiar in the form of the "tone" control on a stereo system or a "graphic equalizer" that has knobs or sliders assigned to each narrow band of frequency.

With noisy speech, one common initial approach is to apply a *bandpass* filter that passes the frequencies of speech while attenuating the portions of the signal with

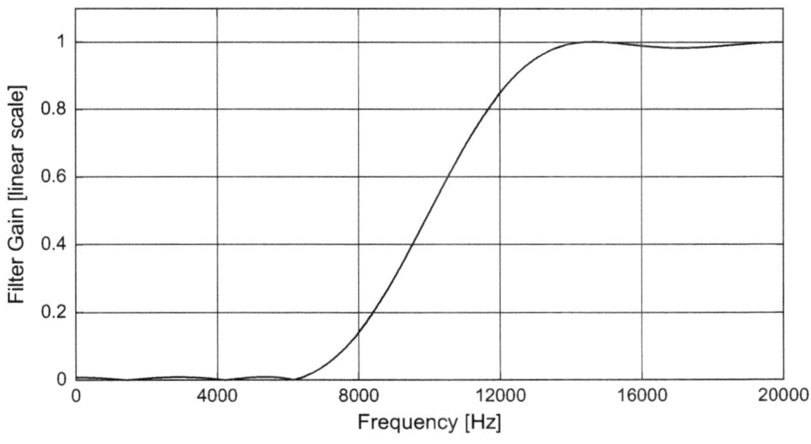

Fig. 6.4 Example spectral characteristic of a highpass filter. Low frequencies are attenuated, and high frequencies are passed through the filter

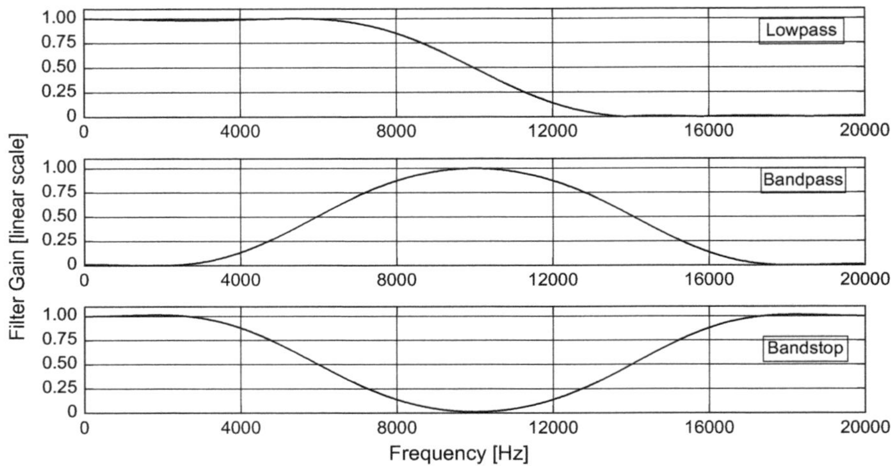

Fig. 6.5 Spectral characteristic of lowpass, bandpass, and bandstop filters

content in frequency ranges lower than or higher than the speech bandwidth. A bandpass filter passing 200 Hz to 4 kHz conveys most of the speech energy needed for reasonable intelligibility while reducing the level of low-frequency rumble and hum. Applying *equalization* to boost the signal slightly in the bandwidth at the upper edge of the speech bandwidth, such as the range 1–4 kHz, can often help emphasize the portion of the spectrum containing the consonant sounds, possibly aiding the intelligibility (Weiss et al. 1974; Weiss and Aschkenasy 1981; Moorer and Berger 1986; Owen 1988).

Speech or other forensic signals containing hum and harmonics corresponding the electrical power system frequency (e.g., 60 Hz and harmonics in North America, 50 Hz and harmonics in Europe) may be improved somewhat using a set of notch filters positioned at the power harmonic frequencies. The hum harmonic frequencies may overlap the desired speech passband, so some care must be taken to balance the reduction in power line noise with the potential degradation in the desired speech signal. Notch filtering may also be attempted for tonal noise at frequencies other than the power system, such as a whine from a ventilation fan, water pump, or some other mechanical system.

Figure 6.6 shows a signal spectrogram in the presence of 60 Hz power line interference. In this example, the first few power harmonics are present (60, 120, 180, 240 Hz). Figure 6.7 shows the signal after a speech bandwidth filter (200 Hz to 4 kHz) and a notch filter with frequencies 60, 120, 180, and 240 Hz. Although still somewhat noisy, the resulting spectrogram and signal show less evidence of the power line noise.

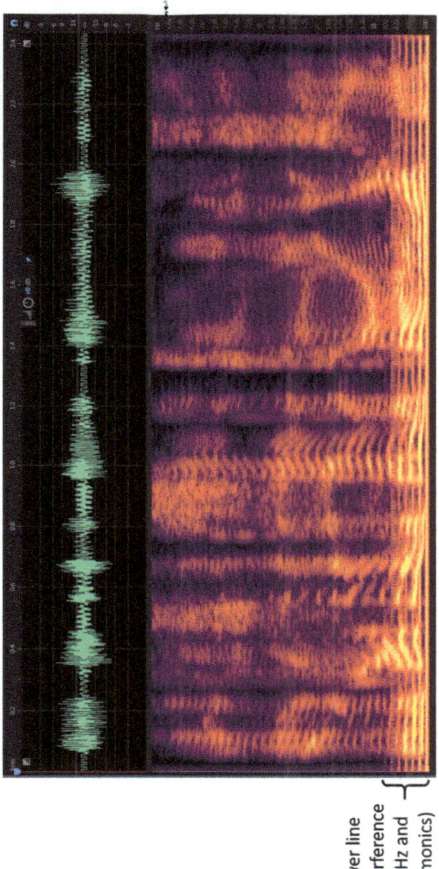

Fig. 6.6 Speech recording with 60 Hz power line "hum" interference (overall duration 2.4 s, frequency range 0–4.8 kHz, linear scale)

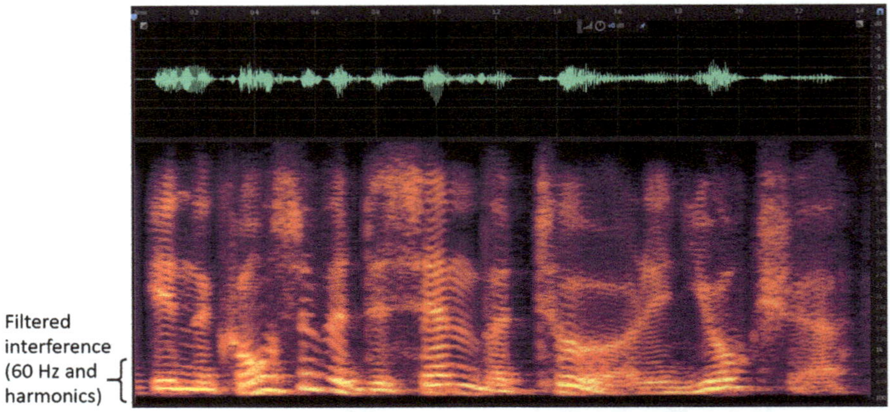

Fig. 6.7 Speech recording filtered with 200 Hz to 4 kHz bandwidth and 60 Hz harmonic notch filters (overall duration 2.4 s, frequency range 0–4.8 kHz, linear scale)

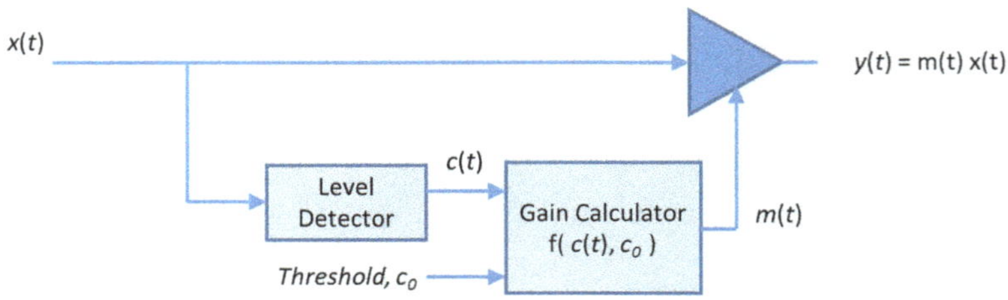

Fig. 6.8 Conceptual diagram of an automatic gain control (AGC) system

6.3.2 Gain Compression and Expansion

While filtering and equalization act primarily upon the frequency content of a recording, other methods are used to adjust the signal level in the time domain. For example, a challenge in many forensic audio recordings is the presence of very loud and very soft passages, such as having one talker very close to the microphone while another talker is some distance away, or a single talker who turns his or her head or moves away from the microphone. For speech recordings, sometimes, there are periods of a few seconds between utterances that are simply distracting background noise.

Some recording systems used for emergency call centers, mobile phones, personal recorders, and surveillance systems may include an automatic gain control (AGC) that detects the short-term level of sound and automatically adjusts the microphone gain so that the loudness stays relatively constant. This process is referred to as *dynamic range compression*, because it attempts to reduce the fluctuation in level over time, compressing the variability (Orfanidis 1996) (Fig. 6.8).

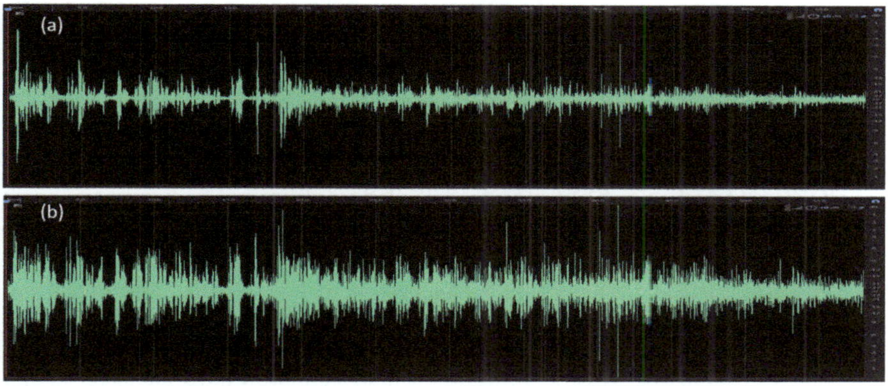

Fig. 6.9 (**a**) Original signal and (**b**) signal with automatic gain control (AGC) set for gain compression. The gain compressor amplifies the quieter passages and reduces the peak levels (overall duration 60 s)

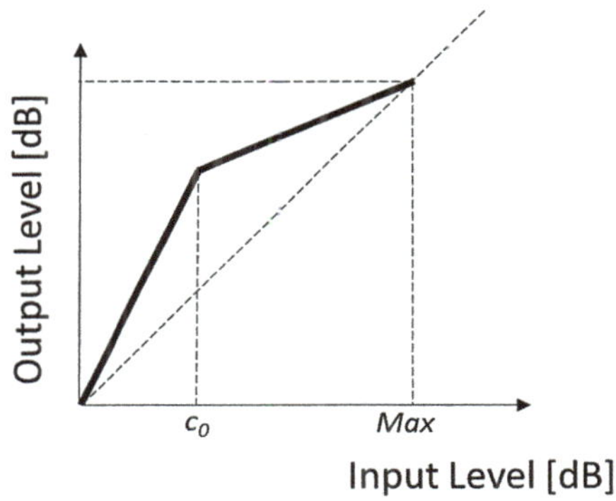

Fig. 6.10 Example output vs. input characteristic for gain compression

The action of an automatic gain control is demonstrated in Fig. 6.9. In this case, the gain calculator compares the level of the signal's envelope compared to a threshold value and boosts the level of the quieter passages relative to the louder passages.

The behavior of a dynamic range compressor is often depicted as an output vs. input level graph, as shown in Fig. 6.10. The graph indicates that as the short-term input level increases up to c_0, the output level is boosted by an increasing amount. As the input level increases above c_0, the output level is boosted by a decreasing amount. The result is that the output level is kept close to the maximum when the input level exceeds c_0.

Systems with an AGC may also incorporate a *dynamic range expander*, frequently called a *noise gate* or *squelch* function. The noise gate has a threshold level below which the gain calculator system automatically decreases or even turns off the input gain under the assumption that if no significant signal is present, the only sounds must be extraneous background noise, and so the volume can be turned down: the gate is "closed" to block the recording of noise. Later, when a louder signal is once again detected at the microphone, the noise gate automatically "opens" and lets the signal through under the assumption that the louder signal is the desired speech. The output level vs. input level depiction for a gain expander is shown in Fig. 6.11, and an audio signal example is shown in Fig. 6.12.

The basic time domain noise gate is made more flexible with careful control of the attack and release times of the gain-controlled amplifier, applying separate gates to two or more frequency bands, and automatically varying the gate threshold to follow the background noise level as it varies with time.

While a forensic audio recording from a system with automatic gain control can be very useful, there may be reasons to be cautious about drawing too many conclusions. For example, the relative loudness of different sounds in a recording may provide clues about the location of different sound sources or movements occurring during the recording, but these clues may be altered if the AGC is active. Similarly, the presence or absence of certain telltale background sounds may provide information useful to an investigation, but the recorder's AGC system may suppress or enhance these background sounds as a side effect of the level adjustments.

Furthermore, AGC systems are not perfect and can be fooled by background noises. The AGC may actually increase the audible noise level during quiet passages between utterances, and there can be what audio recording engineers refer to as audible "pumping" artifacts due to the background noise level increasing and decreasing as the gain changes.

Whether or not the recording system had automatic gain control, the same principle can be applied in the audio forensic enhancement process using a dynamic range processing effect or plug-in for an audio waveform editor. The software dynamic range processor determines the short-term signal level and either boosts or attenuates the signal according to the desired profile. This can help bring out the low-level sounds in the recording without overpowering the louder sounds, but the process may also introduce unintended and undesirable side effects, like boosting noisy portions of the signal. Nevertheless, unlike a real-time processor during the recording, a software dynamic range processing effect can be applied gradually and iteratively to experiment with different settings and adjustments. In some cases, it is

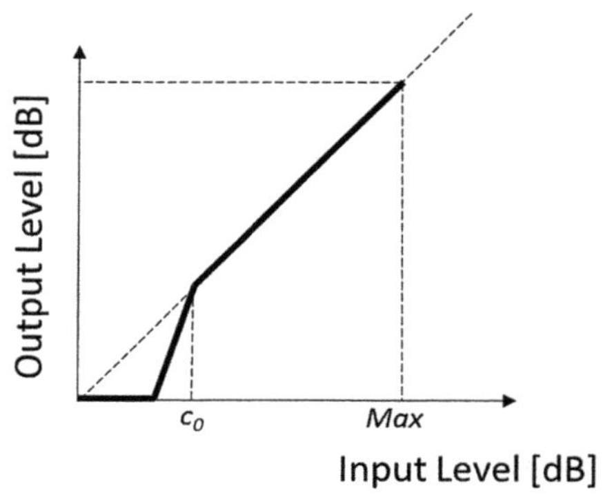

Fig. 6.11 Output level vs. input level for a gain expander

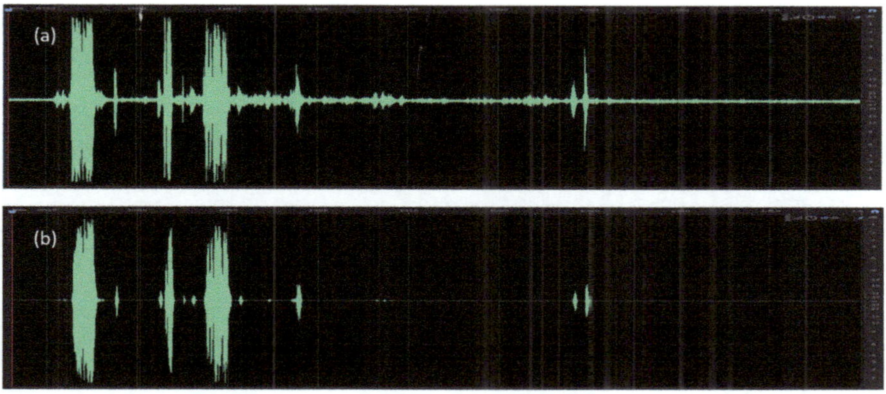

Fig. 6.12 (**a**) Original signal and (**b**) signal with automatic gain control (AGC) set for gain expansion (overall duration 55 s)

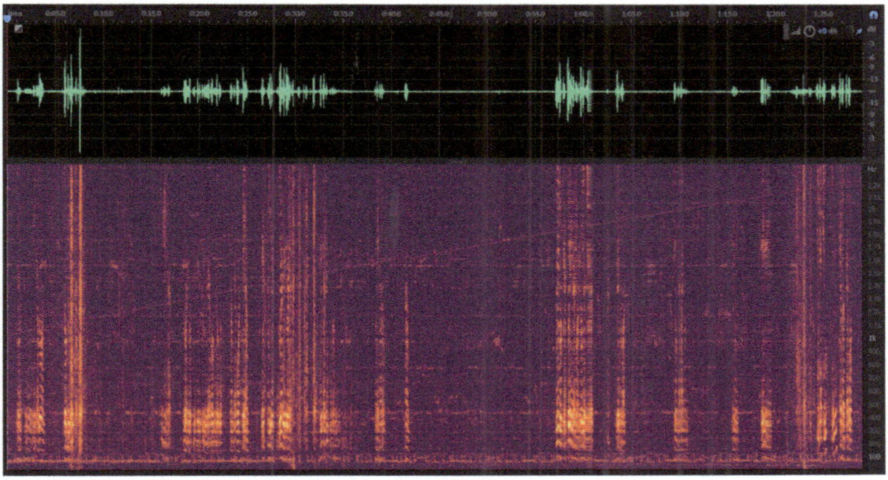

Fig. 6.13 Example forensic recording containing utterances by different individuals with different levels (overall duration 90 s, frequency range 0–2.4 kHz, linear scale)

appropriate to adjust the settings specifically for certain passages of the recording and then use different settings for other portions.

An example speech recording is shown in Fig. 6.13. The recording has several utterances by different individuals at different locations in the room where the recording was made. This example has a noticeable level of background noise and does not include any AGC during the recording. The fluctuating level makes the conversation somewhat difficult to follow, especially when presenting the recording to a review panel or a jury.

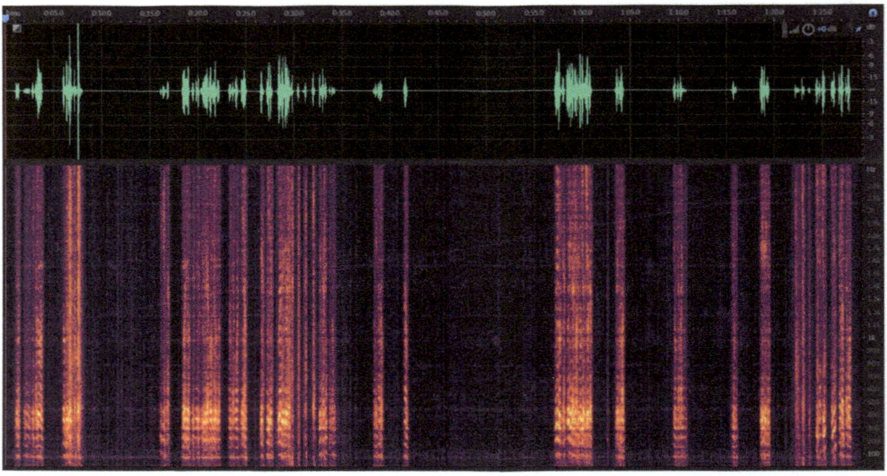

Fig. 6.14 Example recording with gain compression/expansion (overall duration 90 s, frequency range 0–2.4 kHz, linear scale)

Adjusting the gain compression curve so that level is boosted for the low- and mid-level passages, the improved recording is shown in Fig. 6.14.

The action of a dynamic range compressor can be extended to more complicated enhancement structures, including *multiband compression/expansion*. A multiband system separates the signal bandwidth into several overlapping sub-bands and then applies a separate compressor to each one. This technique may be useful in situations involving reverberation, allowing the dynamics to vary across the various resonances that may be present.

6.3.3 Other Important Techniques

Commercial audio forensic software packages typically include a few basic noise reduction settings, including the bandpass and the interference-reduction notch filters described above, as well as some specialized and often proprietary algorithms (Lim and Oppenheim 1979; McAulay and Malpass 1980; Godsill et al. 1998; Koenig et al. 2007). Two common algorithms are *click/pop reduction* and *spectral subtraction*.

An audio click or pop refers to a type of impulsive interference, sometimes referred to as "static," that is caused by electrical or radio frequency interference. Some clicks and pops may be due to loose connections or corroded contacts.

If there are only a few disturbing clicks, it is often possible to reduce the distracting sound by manual editing, as shown in Fig. 6.15a, b. On the other hand, if the recording is long and has many clicks, manual effort may be impractical.

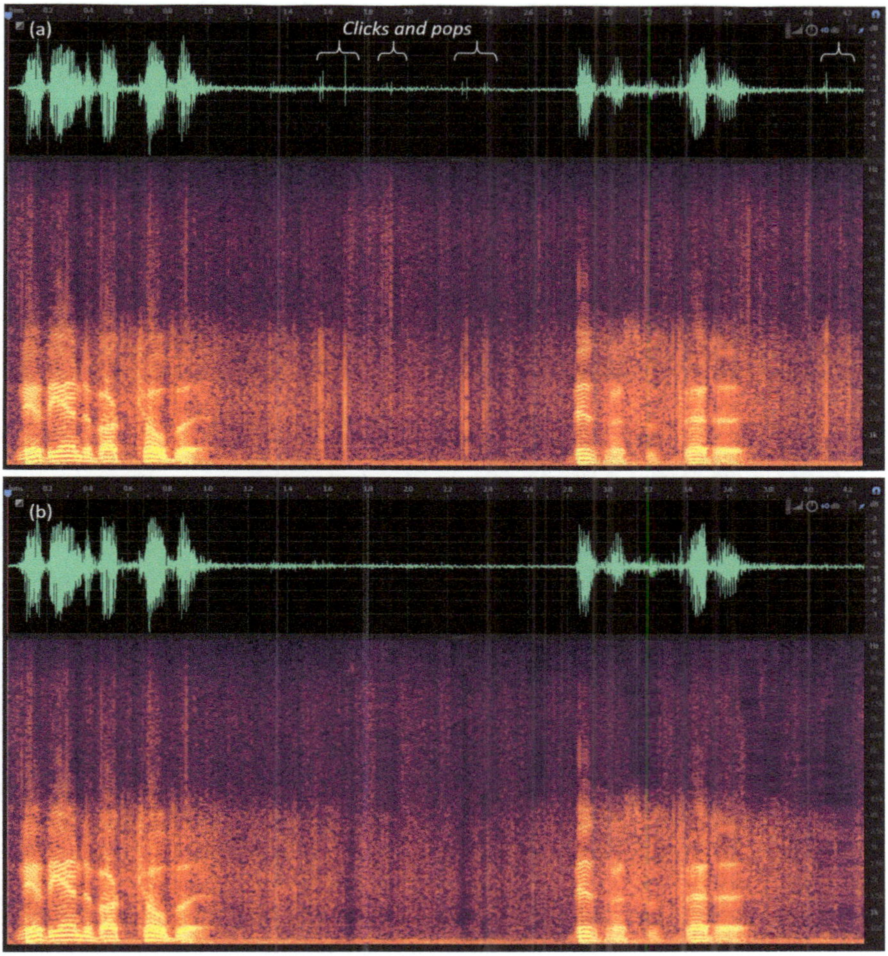

Fig. 6.15 (**a**) Audio recording with distracting clicks and pops and (**b**) signal after click/pop removal by manual editing (overall duration 4.3 s, frequency range 0–10 kHz, linear scale)

Audio forensic software packages and some audio mastering software include algorithms to detect and reduce clicks. The click detection uses a short time window to examine successive segments of the recording seeking abrupt signal changes that would be typical of a click sound. The software provides a choice of the window duration and the level of signal change above which a click is suspected. The software then includes an action to take for the detected clicks, such as reducing the gain so that the click is made less audible. An example of algorithmic click reduction is shown in Fig. 6.16.

It is important to recognize that click detection operates on specific characteristics of the time signal that may also be caused by naturally occurring sounds in the

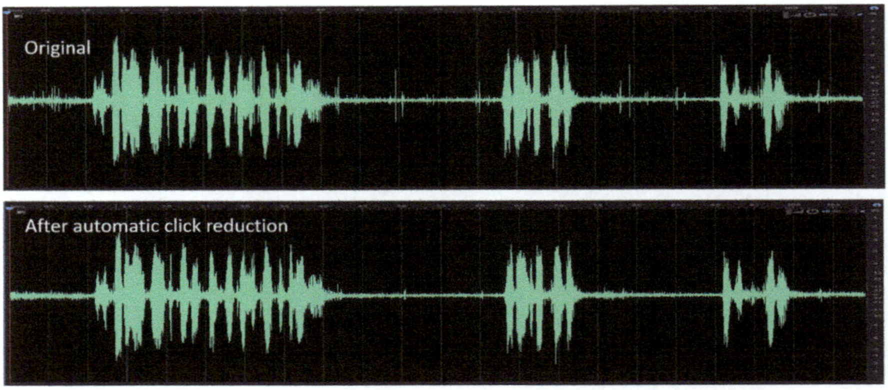

Fig. 6.16 Software detection and reduction of clicks in an audio recording (overall duration 11 s)

original recording. There should be critical listening before and after the click reduction processing to make sure the process does not cause a significant degradation of the signal features (Tsoukalas et al. 1997).

When the background noise is predominantly stationary, *spectral subtraction* can be attempted as a way to improve signal quality (Boll 1979). There are a variety of techniques that fall under the spectral subtraction label, but the basic idea is to measure the background noise characteristics during a portion of the recording in which the foreground signal (such as speech) is absent momentarily. Then, under the assumption that the noise statistics are relatively constant, the noise is modeled using the noise-only portion of the recording, and the modeled noise spectrum can be subtracted from successive short-term blocks on the input signal. The term "subtraction" in this case might better be described as "reduction," because the technique involves the spectral level in a particular frequency range, not the random noise waveform itself.

An example of a noisy audio signal is shown in Fig. 6.17. After processing with spectral subtraction software, the noise-reduced output spectrum is shown in Fig. 6.18. Note that in this particular example, some of the desired signal spectral components are depicted below the noise threshold, so the spectral subtraction process will inadvertently remove them. Nevertheless, the spectral subtraction method may still improve the overall signal-to-noise ratio if the noise level in any particular band is not too high.

Unfortunately, spectral subtraction can sometimes cause undesirable audio artifacts. The audible problems are particularly noticeable when the actual noise level and noise behavior differ from the estimated noise spectrum. The mismatch between the actual noise and the model means that the noise is not subtracted completely, and the residual noise near the model threshold is apparent as a tonal whistling or tinkling sound referred to informally as "musical noise" or "birdie noise" (Cappé 1994).

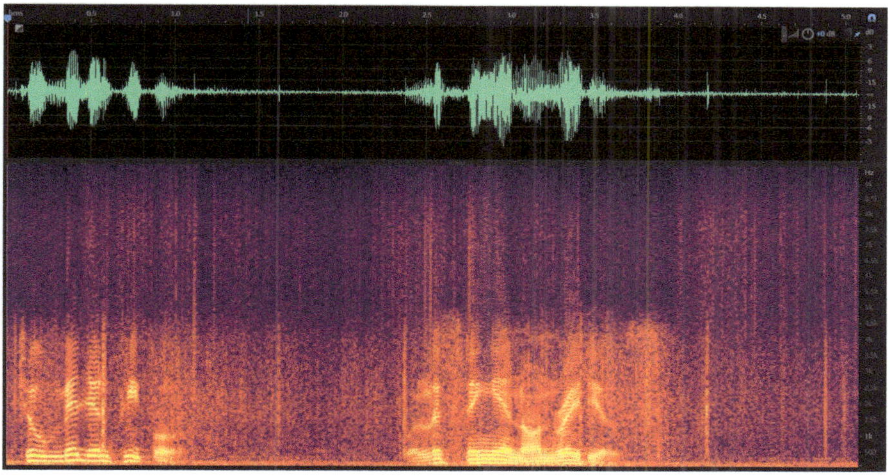

Fig. 6.17 Noisy audio signal example (overall duration 5 s, frequency range 0–10 kHz, linear scale)

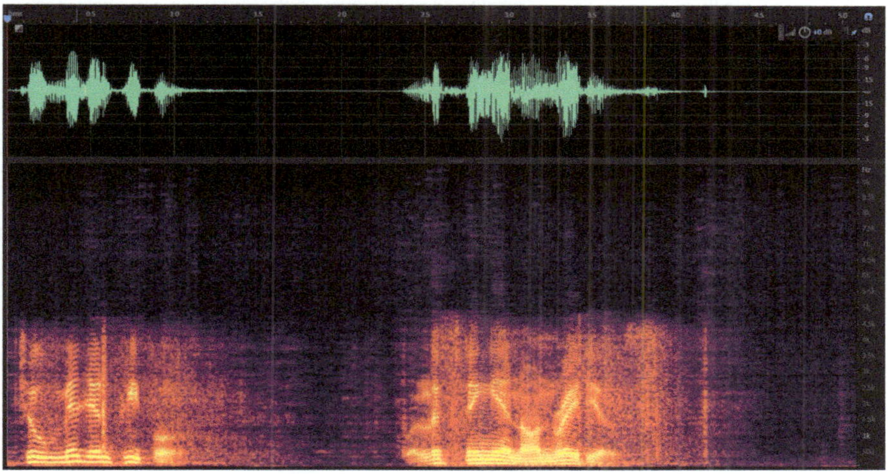

Fig. 6.18 Noisy signal after spectral subtraction processing (overall duration 5 s, frequency range 0–10 kHz, linear scale)

In common forensic enhancement applications, spectral subtraction also requires that the noise level be updated frequently because both the noise spectrum and the desired signal spectrum fluctuate from time to time (Boll 1979). If some of the desired signal's spectral components are below the noise threshold at one instant in time but then peak just above the noise threshold at a later instant in time, the abrupt change in those components may introduce a distracting audible click, pop, or tonal residual.

Therefore, commercial software packages incorporating spectral subtraction rely upon a variety of strategies to reduce the shortcomings. These strategies often include monitoring and automatically updating the noise level estimate, reducing the processing if the desired signal is strong, and incorporating some degree of threshold hysteresis in each frequency band to lessen the appearance of residual musical noise (Maher 2005; Musialik and Hatje 2005).

References

Boll, S. (1979). Suppression of acoustic noise in speech using spectral subtraction. *IEEE Transactions on Acoustics, Speech, and Signal Processing, 29*, 113–120.

Cappé, O. (1994). Elimination of the musical noise phenomenon with the Ephraim and Malah noise suppressor. *IEEE Transactions on Speech and Audio Processing, 2*(2), 345–349.

Godsill, S., Rayner, S. P., & Cappé, O. (1998). Digital audio restoration. In M. Kahrs & K. Brandenburg (Eds.), *Applications of digital signal processing to audio and acoustics*. Dordrecht, The Netherlands: Kluwer Academic Publishers.

Koenig, B. E., Lacey, D. S., & Killion, S. A. (2007). Forensic enhancement of digital audio recordings. *Journal of the Audio Engineering Society, 55*(5), 252–371.

Lim, J. S., & Oppenheim, A. V. (1979). Enhancement and bandwidth compression of noisy speech. *Proceedings of the IEEE, 67*(12), 1586–1604.

Maher, R. C. (2005). Audio enhancement using nonlinear time-frequency filtering. In *Proceedings of Audio Engineering Society 26th Conference, Audio Forensics in the Digital Age, Denver, CO* (pp. 104–112).

McAulay, R., & Malpass, M. (1980). Speech enhancement using a soft-decision noise suppression filter. *IEEE Transactions on Acoustics, Speech, and Signal Processing, 28*, 137–145.

Moorer, J., & Berger, M. (1986). Linear-phase bandsplitting: Theory and applications. *Journal of the Audio Engineering Society, 34*(3), 143–152.

Musialik, C. & Hatje, U. (2005). Frequency-domain processors for efficient removal of noise and unwanted audio events. In *Proceedings of Audio Engineering Society 26th Conference, Audio Forensics in the Digital Age, Denver, CO* (pp. 65–77).

Orfanidis, S. J. (1996). *Introduction to signal processing*. Upper Saddle River, NJ: Prentice Hall.

Owen, T. (1988). Forensic audio and video—Theory and applications. *Journal of the Audio Engineering Society, 36*, 34–40.

Tsoukalas, D. E., Mourjopoulos, J. N., & Kokkinakis, G. (1997). Speech enhancement based on audible noise suppression. *IEEE Transactions on Speech and Audio Processing, 5*(6), 479–514.

Weiss, M. R., & Aschkenasy, E. (1981). Wideband speech enhancement (addition). In *Final Tech. Rep. RADC-TR-81-53, DTIC ADA100462*.

Weiss, M. R., Aschkenasy, E., & Parsons, T.W. (1974). Study and development of the INTEL technique for improving speech intelligibility. In *Nicolet Scientific Corp., Final Rep. NSC-FR/4023*.

Chapter 7
Forensic Interpretation

The basic elements of audio forensics are authentication, enhancement, and interpretation. As noted previously, *authentication* generally involves objective observations and measurements, and like any empirical study, the accuracy and precision of the measurements are very important to assess and to report. Audio *enhancement* tends to be a somewhat more subjective area, as the choices involved in single-ended noise reduction often require a personal judgment about the quality and usefulness of the processed recording. In this chapter, we will learn that forensic *interpretation* needs to be as objective as possible, but the examination leading to the objective interpretation often requires subjective assessment, induction, and experience.

7.1 Scientific Integrity

An important and fundamental principle of forensic science is that the scientific methods and interpretation can be explained in a manner that other experts can understand and with which they will concur. There can be no secrecy of techniques or undisclosed methodology. Especially in criminal proceedings, the findings may ultimately contribute to a court taking away the rights of an individual convicted of a crime or levying a financial judgment or some other sanction. To the extent that scientific analysis of audio evidence serves the needs of the court, the stakes can be very high.

Some audio forensics examiners and expert witnesses received no formal training in the forensic sciences, and these individuals may not be knowledgeable about the necessity of providing complete and verifiable interpretation, including the methods and reliability of the results. Some of these individuals may assert that because they have been involved in many prior cases they can somehow hear things no other expert can detect or explain, or they have developed proprietary techniques

R. C. Maher, *Principles of Forensic Audio Analysis*, Modern Acoustics and Signal Processing, https://doi.org/10.1007/978-3-319-99453-6_7

to analyze recordings that no one else can understand. *It is vitally important that those who choose to enter the fields of forensic science understand that these sorts of assertions are unethical and undermine the integrity of all forensic scientists who appear in court or in other official proceedings.*

In 2009, the US National Academy of Sciences (NAS) published a report entitled *Strengthening Forensic Science in the United States: A Path Forward*. The report was highly critical of the many areas of forensic science, including audio forensics, that have traditionally relied upon subjective analysis and comparison. Among the influential report's comments was the statement:

> Two very important questions should underlie the law's admission of and reliance upon forensic evidence in criminal trials: (1) the extent to which a particular forensic discipline is founded on a reliable scientific methodology that gives it the capacity to accurately analyze evidence and report findings and (2) the extent to which practitioners in a particular forensic discipline rely on human interpretation that could be tainted by error, the threat of bias, or the absence of sound operational procedures and robust performance standards.

Our increasing awareness that subjective forensic findings are not necessarily repeatable from examiner to examiner, nor for the *same* examiner reviewing the same evidence later under different circumstances, heightens our overall concern about forensic subjectivity. These and other similar issues will be considered in the following chapter.

7.2 Methods and Reliability

Scientific measurements have specific *precision* and *accuracy*. While precision and accuracy are often considered synonyms in common parlance, the terms have separate and distinct meaning in science and engineering.

Precision expresses the fine-scale resolution of a particular measurement and its repeatability. A measurement with high precision has more significant digits than a measurement with low precision, which means that proportionately smaller changes to the measured parameter are detectable when the precision is high. Precision also implies the repeatability of a measurement: if the same quantity is observed several times in succession, a precise measurement will return essentially the same measured value each time (low variance).

Accuracy expresses the degree to which the measurement is correct with respect to a known standard reference. An accurate measurement provides a value that is reliably referenced to a *calibration* of the measurement system, and therefore the measurement can be compared to other similarly calibrated measurements.

The types of measurements that arise in audio forensic analysis include time, frequency, amplitude, and spectrum. What are the sources of uncertainty in these measurements? Let us consider the following demonstrative examples and analysis comments.

7.2.1 *Example 1: Simultaneous Recordings from Different Locations*

Consider the following common scenario depicted in Fig. 7.1 Two successive gunshots were recorded by two different dashboard camera systems in law enforcement vehicles parked about 150 meters from the scene of the shooting. Neither camera was pointed toward the shooting scene, but the cabin microphones picked up the sound of the two shots. The position of each vehicle is known, but there is a forensic question about where the shots came from, whether the shots were from the same firearm fired twice in succession, or from two different firearms located several meters apart (Duckworth et al. 1997; Maher 2016). Is it possible to understand the circumstances more fully because two independent recordings are available?

Example forensic questions:

- Position of the first shot with respect to the cruisers?
- Position of the second shot with respect to the cruisers?
- Was it the same firearm from the same position or two different firearms from different positions or one firearm and the shooter moved between shots?

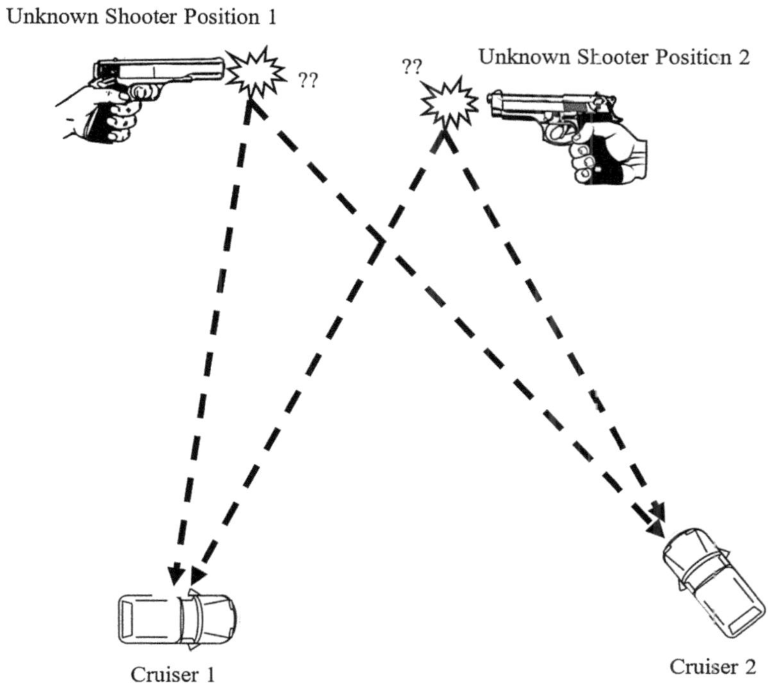

Fig. 7.1 Example scenario in which the sounds of two gunshots are captured by dashboard recorders in two different police cruisers

Measuring time intervals in an audio recording requires identifying two corresponding sound events and then counting the number of audio samples that lie between them. The time interval in seconds is the number of samples divided by the sample rate (samples per second). The uncertainty is often the difficulty in determining the timing of the "two corresponding events," as well as several related issues of uncertainty.

One issue in this scenario is that the two dashboard cameras are not synchronized: The audio recording files begin at different times, and the relative timing of sounds recorded by both camera systems is likely to be ambiguous. Faced with this uncertainty, the examiner needs to determine if there is some way that the two unsynchronized recordings might be synchronized using the audio itself. In the case of dashboard recordings, one possible solution is to identify a signal transmitted from the dispatch center and picked up simultaneously by the radios in the two cruisers: the cabin microphone in each dashboard recorder would capture the dispatcher's speech. The audible signal common to both cruisers is used to align the two unsynchronized recordings. If no dispatcher radio signal is available, the examiner could attempt to find another common signal from a known location at the scene and arrange to use that signal for time alignment.

Another potential issue in this scenario is the uncertainty about the precise location of the vehicles. The general positions and orientations of the vehicles are likely to be estimated from the video or perhaps still photographs of the scene, but the accuracy of the location determination is certainly limited. Any interpretation requiring geometric calculations will depend upon the position uncertainty. Sound in air travels approximately 1 ft per millisecond, so position uncertainty on the order of a meter or two would result in acoustical timing uncertainty of several milliseconds. The examiner would need to include the position uncertainty when performing and reporting upon the associated calculations.

Along with the first-order timing and position uncertainty, the examiner would also need to review a second-order consideration: the air temperature. The speed of sound in air increases proportionally to the square root of absolute temperature. For example, the speed is 331 m/s at 0° C, 343 m/s at room temperature, and 353 m/s at normal body temperature. Any uncertainty about the temperature leads to uncertainty in the sound speed, which in turn leads to uncertainty in distance based on the elapsed time between events. This uncertainty may be small if the temperature is estimated reasonably well, but any related assumptions need to be checked before reporting the ultimate findings.

Assuming the examiner can synchronize the two recordings and fix the positions of the two cruisers, it may be feasible to use the resulting audio to address the forensic questions about the position of the shots. Identifying a sound source location based upon simultaneous observations at two known positions uses a calculation procedure known as *multilateration*. Multilateration is based upon the *time difference of arrival* (TDOA) of a sound at two or more receiver positions. The TDOA multilateration approach is sometimes erroneously referred to as "triangulation," which is a different procedure using angle measurements relative to known positions.

The principle of multilateration is that an impulsive sound produced by a source will propagate in all directions at the speed of sound, arriving at the receivers with a time delay corresponding to the source-to-receiver distance divided by the speed of sound. Ordinarily, the synchronized receivers do not know the absolute time that sound was produced from the unknown source location, but if synchronized, they do know the *time difference* of arrival. Using the relative time difference, Δt, and the speed of sound, c, the source must be at a location that is a distance $L - c\ \Delta t/2$ from the receiver that first receives the signal and a distance $L + c\ \Delta t/2$ from the second receiver. If we assume that the source and the two receivers are all in the same plane, the locus of possible source points satisfying the relative distance constraint is a mathematical *hyperbola*, given by the equation

$$\frac{x^2}{x_a^2} - \frac{y^2}{\left(x_0^2 - x_a^2\right)} = 1,$$

where we assume a Cartesian (rectangular) coordinate system centered between the two receivers located at positions $(-x_0, 0)$ and $(+x_0, 0)$, coordinate x_a corresponds to $c\ \Delta t/2$, and the source at unknown location (x, y). This configuration is shown in Fig. 7.2. The depiction in the figure is for the case in which the source signal arrives at Receiver_1 Δt s before it arrives at Receiver_2.

Using this formulation, the audio forensic examiner would use knowledge of the receiver (cruiser) positions and the TDOA (Δt) to estimate the hyperbolas of possible shot locations for the first shot and the second shot. Assuming the cruisers did not move, and the Δt is different for the two shots, the examiner could conclude that

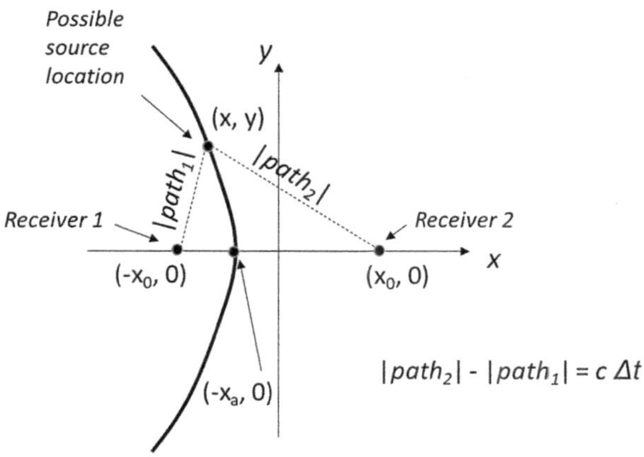

Fig. 7.2 Geometrical configuration for two-receiver multilateration. The coordinate system is chosen so that the x-axis is a line connecting the two receivers, and the y-axis intersects half way between the receivers. This figure assumes that the source is at some location (x, y) such that a pulse it produces arrives at Receiver_1 Δt s before it arrives at Receiver_2. The locus of points (x, y) that meet this condition describe a mathematical hyperbola with vertex $|x_a| = c\ \Delta t/2$

the shots came from different locations, but the two-receiver formulation does not pin down the precise locations, just the locus of possible locations.

The question of whether the two shots were from two separate firearms, or possibly from a single firearm that moved between the shots, would need additional information about the time between shots and the likely distance a shooter could move during that time interval. Thus, the audio forensic examiner would need to provide the acoustical findings to the investigator, who then might be able to combine the multilateration results with other information from the incident investigation.

What if another synchronized recording was available from, say, a third cruiser at another nearby location? In that case, it may be possible to perform *two* multilateration calculations using pairs of receivers, i.e., determine the hyperbola for Receiver_1 and Receiver_2 and then determine a second hyperbola for Receiver_1 and Receiver_3. Assuming the two hyperbolas intersect at one or more points, the results can provide location estimates for the shot(s).

In practical cases, the uncertainties of receiver position, receiver synchronization, and detection of TDOA may be sufficient to give significant uncertainty in the location estimates. The examiner needs to take into account and fully explain all of the potential sources of error or discrepancy when preparing the case report.

7.2.2 *Example 2: Recording Involving Doppler and Converting to Speed*

For a second example scenario, a 911 call center records the telephone dialog between a caller and a dispatcher. The caller explains that his vehicle is stalled and pulled over in the breakdown lane (Fig. 7.3). While the dispatcher is collecting information and calling the emergency responders, the sound of an approaching truck's air horn is heard in the background. A second later, there is a crash, and the telephone call ends abruptly. Later, during the investigation, the call recording is analyzed, and the frequency of the truck's horn in the recording is determined to be 329 Hz.

The nominal frequency of the air horn used in the particular truck is 295 Hz, which was verified by testing the horn after the accident. The forensic question:

Fig. 7.3 Stalled car and an approaching truck sounding its horn

- What was the estimated speed of the approaching truck, based upon the Doppler effect?

Due to the Doppler effect, a sound source moving at speed v directly toward a stationary receiver results in a higher frequency being observed. The received frequency is $f_o\ (c/(c - v))$, where f_o is the sound frequency produced by the source at rest, and c is the speed of sound in the air. Assuming the air temperature, the source frequency, and the received frequency are known, the source motion with respect to the stationary receiver can be calculated: $v = c\ (1 - f_o/f)$.

In this case, the air temperature is known to be approximately $T_C = 17°$ C (63° F), corresponding to a speed of sound $c = 331.3 * \text{sqrt}(1 + T_C/273) = 341.5$ m/s. Then with $f = 329$ Hz, and $f_o = 295$ Hz, this gives $v = 35.3$ m/s (78.9 miles per hour). So based on the audio evidence, the truck was approaching at nearly 80 miles per hour before the crash.

There are several areas of uncertainty in this scenario. First, the determination of the measured 329 Hz signal frequency depends upon the method used to estimate the frequency from the audio recording, any uncertainty about the precise sampling rate of the signal, and so forth. Second, the air temperature may not actually be known with great precision. Finally, if the approaching truck was not traveling directly toward the receiver, the Doppler calculation could underestimate the truck's speed, since the Doppler frequency reflects the radial velocity component with respect to the microphone.

The audio forensic examiner should be able to verify that the audio recording's sampling rate is correct and that the frequency determination technique is reliable, by generating a precise test tone and making a recording with the same system originally used by the 911 call center. The reason that the examiner needs to verify these and other parameters is based upon a sense of professional integrity—and wise paranoia. It is often expedient to assume that recording details, timing, fiduciary markings, and so forth are all correct, but there have been situations in which an audio recording stated to be made with an 8 kHz sampling rate was actually made with a nonstandard rate of 8192 Hz. The sampling rate error of 2.4% means that a tone recorded at 8192 Hz sampling rate and then played back at an 8000 Hz rate will be out of tune (flat) by about half a semitone. A discrepancy like that in the Doppler calculation would result in a frequency of 321 Hz instead of 329 Hz, and this would indicate a truck speed of 27.7 m/s (62 miles per hour)—significantly slower than the 78.9 miles per hour estimate.

Also, the examiner should calculate any variability due to an incorrect air temperature. In this case, varying the air temperature over the range 15–20° C (59–68° F) results in only a small variation in the range of the vehicle speed estimate: 35.2–35.5 m/s (78.7–79.4 miles per hour). This audio forensic information, combined with other physical evidence the investigators may have, can provide useful information to help understand the accident.

7.2.3 *Example 3: Sound Level Vs. Distance*

A third illustrative audio recording scenario is depicted in Fig. 7.4. One individual with a memo recorder in a shirt pocket recorded a shouted utterance of a threat made by a second individual outdoors on a suburban street. Both individuals agree that the recording is authentic, but there is a dispute regarding the distance between the individual alleged to have made the threatening statement and the individual with the recording device. The threatened individual claims that the shouting took place only 1 m away (Scenario 1) in a very menacing manner, while the individual accused of shouting the threats claims to have been more than 8 m away across a street (Scenario 2) and not physically menacing. No other witnesses or physical evidence is available.

The forensic question:

- Can the distance between the shouting person and the recorder be estimated from the audio recording itself?

It is understood that the sound level from a source is expected to decrease with increasing distance due to wave physics. Neglecting reflections and reverberation, the sound intensity follows a spherical spreading pattern, resulting in the sound pressure amplitude being proportional to 1/distance. In terms of sound pressure level (SPL), this means that the level decreases by 6 dB for every doubling in distance.

In a reverberant environment, some sound reaches the microphone after reflecting off nearby objects and surfaces, so the level may actually be somewhat higher than predicted by the spherical spreading model. The relative balance between the

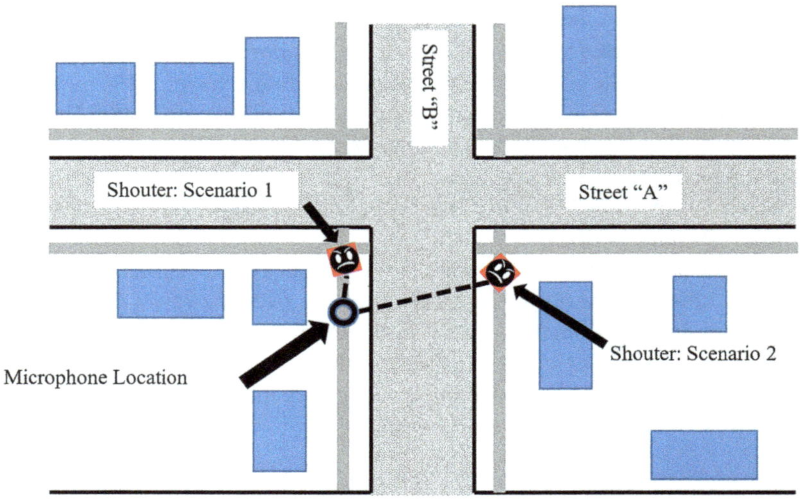

Fig. 7.4 Plan view of possible scenarios for threatening statement

direct sound and the reflected/reverberant sound will also vary if the source is closer to the microphone or farther away. When close, the recorded sound will be dominated by the direct sound of the shouter, while if farther away, the direct and reflected sound will be of a comparable level. In any case, it is expected that a sound recorder will have a clearly different recorded level for someone shouting 1 m away compared to the same individual shouting from 8 m away. The spherical spreading model would indicate a sound pressure level difference of 18 dB, which would be very noticeable.

Nevertheless, there are several areas of uncertainty in this example. First, the loudness of someone shouting is not a calibrated source, so the SPL cannot be known precisely. Second, the location and sensitivity of the recording microphone can have a substantial effect upon the recorded signal. Furthermore, many "memo recorder" devices include automatic gain control (AGC) electronics that boost the signal automatically if it is too quiet and attenuate it automatically if it is too loud. This means that the recorded level from a nearby source and from the same source when moved farther away may actually end up being similar. Finally, as noted above, the effects of the recording environment may have a significant impact on the signal due to sound reflections/echoes, reverberation, and ambient background noise. In some cases, the presence of distinct sound reflections may help determine some details about the position of the source and microphone.

Faced with these areas of uncertainty and the basic forensic question involved, an examiner in this case should probably request to do a reconstruction of the recording circumstances. The investigator could provide the memo recorder used, and the audio forensic examiner could determine the appropriate recorder settings used to obtain the original evidentiary recording. Then a test could be done at the actual scene of the incident, reconstructing the circumstances of the disputed scenarios. In this sort of reconstruction, the examiner would be wise to make a complete and uninterrupted video recording of the scene to help document the test procedures and minimize any challenge to the validity of the experiment.

7.3 Likelihood Ratios

The importance and reliability of forensic evidence depends upon a variety of factors in an investigation. Audio forensic evidence is usually interpreted with a combination of objective and subjective considerations, and some level of uncertainty is nearly always present. While a scientific study can give an indication of the uncertainty, and ongoing analysis may provide additional insights in the future, forensic examination is not usually subject to ongoing review. The court needs to make a judgment at the time the case is heard, so the legal authorities need to weigh the various pieces of evidence and assess whatever level of doubt there may be (Morrison 2011).

As noted in the National Academy of Sciences report on forensic science (2009), the US legal system increasingly accepts the statistically strong likelihood provided by DNA evidence as the ideal standard for all forms of forensic evidence. DNA results use a *likelihood ratio* (LR) formulation to state the results of a DNA comparison. The LR is a way to express the probability of a certain observation given two competing hypotheses (Perlin 2010; Lindley 2014). In the context of a criminal case, the LR is the ratio of two probabilities: the probability of an observation being made given that the suspect is, in fact, the perpetrator divided by the probability of the same observation being made but the perpetrator was someone other than the suspect on trial. The first probability (numerator) is referred to as the probability of the *prosecution hypothesis*, because that is the prosecutor's theory of the crime. The second probability (denominator) is referred to as the *defense hypothesis*, since it represents the defense argument that the suspect is innocent.

With DNA evidence, the statistical likelihood of securing a pattern match between a suspect's DNA and a sample of DNA evidence is currently considered to be exceedingly selective. Statements such as "*a match between the stained garment and the defendant is 9 quadrillion times more probable than a coincidental match to an unrelated person*" are based on the likelihood ratio calculation. The LR in this case has a numerator of 1, since it is assumed that the DNA markers would match if the suspect contributed the DNA with probability 1. The denominator would be 1 over 9 quadrillion (the probability of a random DNA pattern taken from the sample population would match the same number of markers).

Of course, a DNA "match" is not necessarily an automatic determination of guilt. The match may be caused by a contaminated sample or by some other valid—and innocent—reason that the suspect's DNA might have been present at the crime scene other than due to committing the crime. Therefore, the prosecution, defense, and the judge still have to interpret all of the available evidence carefully and completely.

Is it feasible to use a likelihood ratio procedure with evidence other than DNA, such as audio forensic evidence? The answer is a qualified yes, but the difficulty often lies in assessing the probabilities corresponding to the prosecution and defense hypotheses. The extracted markers and mathematical arguments are quite different from the DNA procedures.

For example, audio forensic evidence might include the recorded utterance of an individual that the prosecution claims is the defendant in the case, while the defense may argue that the defendant did not speak the words present in the recording.

In order to present this audio forensic evidence in a meaningful way, the *selectivity* of the voice matching process needs to be established. This is not easy, because the attempt to identify a talker will depend upon the length of the recording, the signal quality, how many words are uttered, and a variety of subjective considerations. Even when a talker *tries* to utter an example phrase exactly the same way two times in a row, the speech signals will be different in many details. This is quite different from a DNA test, which is expected to give exactly the same markers on every test.

Moreover, the probability of the prosecutor's hypothesis—that the defendant uttered the recorded words—may be difficult to establish because this would require many example segments spoken by the defendant under circumstances identical to the evidentiary recording, and a test to see how many of the known utterances were properly identified to assess the reliability.

Similarly, the probability of the defense hypothesis—that someone other than the defendant uttered the recorded words—would require statistical knowledge of the population of other individuals who speak the same language and are otherwise similar to the defendant (age, gender, accent, etc.), and the likelihood that one of those similar talkers would be classified as matching the recorded speech.

For these reasons, it has been uncommon to see the likelihood ratio presented in audio forensic cases such as talker identification. This continues to be an area for current and future research.

References

Duckworth, G. L., Gilbert, D. C., & Barger, J. E. (1997). Acoustic counter-sniper system. In E. M. Carapezza & D. Spector (Eds.), *Proc. of SPIE, command, control, communications, and intelligence systems for law enforcement* (Vol. 2938, pp. 262–275).

Lindley, D. V. (2014). *Understanding uncertainty*. Hoboken, NJ: Wiley.

Maher, R. C. (2016). Gunshot recordings from a criminal incident: Who shot first? The Journal of the Acoustical Society of America *139*(4), 2024. Lay-language version: http://acoustics.org/2aaaa7-gunshot-recordings-from-a-criminal-incident-who-shot-first-robert-c-maher/

Morrison, G. S. (2011). Measuring the validity and reliability of forensic likelihood-ratio systems. *Science & Justice, 51*, 91–98.

National Academy of Sciences. (2009). *Strengthening forensic science in the United States: A path forward*. Washington, DC: National Academy Press.

Perlin, M. W. (2010). Explaining the likelihood ratio in DNA mixture interpretation. In *Proceedings of Promega's Twenty First International Symposium on Human Identification, Madison, WI.*

Chapter 8
Expert Reports and Testimony

Following the forensic examination, the "product" is often a formal report. Depending upon the nature of the investigation, the report may be a brief description of the techniques used and the measurement results, or it may be a more extensive document with figures, photographs, data tables, and scientific conclusions. If the report is intended for use in a court of law, the examiner may need to be qualified as an "expert" before the report can be admitted as formal testimony.

8.1 Qualification as an Expert

What does it take to be considered an expert for audio forensic testimony? The attorneys involved in the case will guide the expert qualification process, but it is helpful to understand the common questions the audio forensic examiner will be asked.

In the United States, the way in which the court determines admissible expertise varies from one jurisdiction to another. The standards may be based upon the case *Frye v. United States*, 54 App. D.C. 46, 293F.1013, DC Ct App 1923 (referred to as "Frye"), the "Daubert" case, *Daubert v. Merrell Dow Pharmaceuticals*, 509 U.S. 579 (Supreme Court of the U.S. 1993), or some similar standard.

The Frye standard, in simple terms, requires that the methods and techniques of the expert be generally accepted as reliable in the relevant scientific community. Several states use the Frye standard to determine admissibility of expert testimony.

The Daubert standard uses the Federal Rules of Evidence to test the relevance and scientific reliability of the expert's testimony. Subsequent cases, such as *Kumho Tire Co. v. Carmichael* (526 U.S. 137 1999), have extended the application of Daubert standards to subjects beyond purely scientific testimony, such as technical fields and engineering principles including audio engineering. Expert testimony in the US Federal Courts is tested for admissibility using the Daubert standard, and also many states use Daubert as the standard for establishing expertise.

R. C. Maher, *Principles of Forensic Audio Analysis*, Modern Acoustics and Signal Processing, https://doi.org/10.1007/978-3-319-99453-6_8

In 2011, Rule 702 of the Federal Rules of Evidence was amended to read:

RULE 702. TESTIMONY BY EXPERT WITNESSES

A witness who is qualified as an expert by knowledge, skill, experience, training, or education may testify in the form of an opinion or otherwise if:

(a) The expert's scientific, technical, or other specialized knowledge will help the trier of fact to understand the evidence or to determine a fact in issue;
(b) The testimony is based on sufficient facts or data;
(c) The testimony is the product of reliable principles and methods; and
(d) The expert has reliably applied the principles and methods to the facts of the case.

The court may apply the Daubert standard using a set of questions for the examiner. The questions typically include the following areas:

1. *Reliability*: whether the expert's technique or theory can be tested in some objective sense or whether it is instead simply a subjective opinion that cannot be assessed in a systematic manner
2. *Review*: whether the technique or theory has been subject to peer review and publication
3. *Uncertainty*: the known or potential rate of error of the technique or theory when applied
4. *Standards*: the existence and maintenance of standards and controls
5. *Reputation*: whether the technique or theory has been generally accepted in the scientific community

The expert qualification process points toward the use of *accepted* and *objective* techniques in audio forensic analysis. An expert's reliance upon proprietary techniques that cannot be assessed or upon subjective impressions and assertions without objective basis is a red flag in any area of forensic studies. As noted in the highly critical 2009 report on forensic examination from the National Academy of Sciences, there is growing judicial awareness and concern about the reliability of expert testimony and an increasing expectation that disreputable or purely subjective opinions will not be admitted for expert testimony.

While an audio forensics expert might think he or she could secure future forensic analysis assignments more easily by developing proprietary analysis techniques that no other examiner has, it is essential to remember that modern courtroom testimony must be fully understandable by the court, not veiled in secrets or unsubstantiated assertions. All techniques and expert opinions must be based upon techniques that can be explained, that have a known level of uncertainty, and that are part of the standard analytical repertoire of acoustical science and audio engineering.

In particular, expert statements that use "golden ears" assertions, such as "*Because of my superior experience and keen hearing, I can perceive something that other people cannot hear*," are to be eschewed. Expert statements have to be explained in a manner that can be confirmed by other experts and understood by the court.

8.2 The Expert Report

After performing the requested audio analysis assignment, the audio forensic examiner may be asked to prepare a formal report. The format and contents of the report may vary depending upon the needs of the investigator who requested the forensic examination, but the common outline will include the following elements:

I. Cover page
   A simple sheet listing the report title, date, case identification, examiner name and address, and similar basic information.
II. Introduction
   The introduction provides the context and circumstances for the forensic examination, the nature of the evidence, and the reason for the audio forensic examination.
III. Summary of case facts
   Following the introduction, the case summary gives more detail about the facts presented to the examiner regarding the source of the audio material, the date, time, and location of the recording, any initial concerns regarding the quality and integrity of the material, and the outline of the examination steps carried out.
IV. Summary of qualifications
   It is customary in an expert report to include a summary of experience and qualifications. A full curriculum vitae (professional resume) is included as an appendix to the report, so the summary of qualifications need only give the examiner's formal education; years of experience as an audio forensic examiner; any significant publications, awards, and certifications; and a description of the examiner's current and previous professional positions.
V. Initial examination of the audio material and its integrity
   This section is generally short and contains a description of the manner in which the audio evidence was received and verified in the examiner's lab. As indicated previously, the examiner will make a bit-accurate digital copy of the evidence for processing, leaving the original file unaltered.
VI. Audio forensic questions
   This section lists the specific audio forensic questions addressed in the analysis. For example, "When is the first gunshot sound detected in the first 135 s of the recording?" or "Is it possible to increase the intelligibility of the utterances between 10 m 45 s and 11 m 30 s in the recording?"
VII. Explanation of evidence
   In the context of the case and the audio forensic questions, this section of the report includes a description of the audio recordings, the location of the recording microphones, the type and known or presumed characteristics of the recording system, and other parameters associated with the audio evidence.
VIII. Explanation of techniques

Answering the audio forensic questions typically requires waveform analysis, spectrographic analysis, critical listening, and so forth, as explained in Chap. 5. The techniques section of the report describes the standard methods and procedures used in the analysis. The description needs to be of sufficient detail that another expert skilled in audio forensic analysis can understand and replicate the steps based upon the steps described in this section.

IX. Findings

The concluding section states the findings and conclusions, typically expressed as answers to the specific audio forensic questions presented in Section VI of the report.

X. Appendix

Curriculum vitae, supplementary graphical information, and other important supporting information

The audio forensic findings expressed in Section IX of the report may be of critical importance to the overall investigation. The investigators, attorneys, and/or the court will typically combine the audio evidence with the available physical evidence, forensic studies, witness statements, and other circumstances of the incident. Therefore, it is vital that the report provides information about the reliability, potential sources of error, and level of uncertainty in the findings.

As noted in Sect. 8.1, it is not appropriate for the examiner to include unsupportable assertions and statements in the report. Statements like "in my experience, I can tell by listening that the recorded sound is from a 38 caliber handgun" or "based upon more than a dozen similar cases I've worked on, the voice in the recording has to be the utterances of the defendant" are not acceptable. If the examiner has a "hunch" or a subjective opinion about the evidence, the subjective information must be supported with facts and objective interpretation, or it should not be included in the report.

Despite the call for stronger objectivity and assessment presented in the National Academy of Sciences 2009 report on forensic sciences, in some jurisdictions it is still assumed that an expert's *opinion* can be given greater credence simply because he or she has extensive experience in the field of interest. In these jurisdictions, it is customary that the expert precede the statement of any specific findings with the assertion: "*I hold the following opinions to a reasonable degree of scientific certainty in the field of forensic audio analysis*."

While the phrase *reasonable degree of scientific certainty* may be requested for forensic expert reports and might seem to a nonprofessional to imply a meaningful standard of reliability, those peculiar words are never used in a regular scientific context such as a scholarly journal, research seminar, or a formal scientific research report. Scientific peer reviews are based upon a methodology and an established level of statistical uncertainty, not a "reasonable degree" of certainty.

When an audio forensic examiner uses the "reasonable degree of scientific certainty" statement in an expert report, it is considered good practice nowadays to provide a more meaningful scientific justification for the related findings, not just an implicit assertion of "trust me, I'm an expert."

8.3 Expert Testimony

A case involving audio forensic evidence may sometimes end up in a courtroom or some other official venue that requires a personal appearance by the examiner. This could be a criminal case involving a defendant accused of a crime, a civil lawsuit of some kind, or an investigation into an accident or some other incident. If the case is built upon the audio evidence and its interpretation, the audio forensic examiner may be called to testify in a deposition, in a public hearing, or in a court of law.

The admissibility of expert testimony in court follows one of the standards mentioned above: *Frye*, *Daubert*, or a state judicial standard. If there is a dispute over the expert's methods and knowledge, the opposing legal counsel may challenge the expert's appearance.

8.3.1 The Role of the Expert

As has been mentioned several times, no matter which side has hired the audio forensic expert, the expert is *not an advocate* for the guilt or innocence of the defendant. The expert's role is to:

(a) Help explain to the court the facts, technology, methods, and acquisition of the audio evidence.
(b) Give a concise summary of the relevant acoustical and physical principles needed to understand the evidence.
(c) Explain what can and what cannot be determined scientifically, given the specific kinds of evidence available in the case.
(d) Provide an explanation of the analysis performed; the methods used; the reliability, accuracy, and tolerance of those methods; and the basis for the specific conclusions drawn.

In a court case, it is the duty of *the attorneys*, not the expert witnesses, to take all of the evidence and testimony and make the legal arguments for the jury and/or judge on behalf of their clients.

8.3.2 Deposition

A common legal procedure in the United States is a *deposition*. A deposition is a formal question-and-answer session in which the opposing attorney asks questions of the witness. The deposition is taken under oath, and a professional court reporter (stenographer) and a videographer typically make a legal record of the proceedings. Although depositions rarely take place in criminal cases, they are very common in civil litigation.

While a *fact witness* provides factual observations without offering opinions, an *expert witness* (such as an audio forensics examiner) generally answers questions about a previously produced expert report and offers professional oral opinions in response to the opposing attorney's questions. As noted previously, an expert *opinion* is more accurately described as a testable scientific statement based upon standard techniques and bona fide evidence, rather than the more customary and informal interpretation of the word "opinion."

The deposition provides the opposing counsel the opportunity to examine the witness in advance of the trial, and since the testimony is under oath, the information from the deposition can become a key part of the subsequent legal proceedings. Depositions in the United States follow procedural rules determined by the relevant court, but they are *adversarial* events coordinated by the opposing attorneys in the case. Among other things, this means that statements elicited under oath during a deposition can be used by the opposing counsel to prepare a strong cross-examination during a subsequent trial. Therefore, it is essential to be prepared properly prior to providing the deposition testimony. The attorney who hired the expert will provide the preparation and advice prior to the deposition itself. The attorney can voice objections to the comments of the opposing counsel during the deposition, but because there is seldom a judge present at the session, the objections are simply entered into the transcript and the questioning continues.

8.3.3 Testimony and Demeanor

Audio forensic deposition and courtroom testimony usually revolves around the introduction of the audio evidence to be used by the parties in the case. The expert helps explain the circumstances and characteristics of the audio recording and describes the analysis procedures and findings.

Although most attorneys, judges, and jury members these days will have general familiarity with sound recordings and audio concepts, few will have knowledge and experience in any way comparable to the expert. Therefore, the attorney who calls the audio forensic expert may choose to use a portion of the testimony to introduce the education, training, experience, and other general qualifications of individuals who serve as audio forensic examiners and then discuss the specific qualifications of the expert witness. The attorney may also ask the witness several questions that comprise a brief tutorial about the key scientific principles needed to understand the evidence and testimony in the case. This process becomes mandatory if the examiner's methods and qualifications are disputed by the opposing counsel: The court may require a *Daubert hearing* (see Sect. 8.1) to consider if the expert's testimony is admissible.

8.3.4 Cross-Examination

After the expert provides direct testimony at the request of the lawyer who hired the expert, the opposing counsel has the opportunity for *cross-examination*: to ask questions about the issues raised in the direct testimony, often in an attempt to call into question the expert's techniques and interpretation.

The attorney who has hired the expert will provide pretrial preparation to help anticipate the kinds of questions that are likely during the cross-examination portion of the testimony. It is important for the expert to remember that the opposing counsel providing the cross-examination has the duty to raise doubts about the issues brought up during the direct testimony in a manner that best serves the interests of their side of the case. This means that the opposing counsel may appear argumentative and skeptical or jovial and friendly or in whatever demeanor the attorney decides will best advance their theory of the case.

Despite the fact that the expert is probably the most knowledgeable person in the courtroom regarding the audio evidence in the case, the expert's deportment must never be condescending or argumentative, even if the opposing counsel presents a challenging sequence of questions or interruptions. Furthermore, the expert must remember that the proper role is *to testify* about the evidence and educate the court regarding the science, *not to argue the case*—which is the lawyer's role. The expert must not attempt to debate, quarrel, or "outsmart" the opposing counsel.

In short, the "prime directive" of expert testimony is to testify truthfully. While this may seem obvious, the fact that the judicial process is an adversarial system tends to exaggerate the peculiar interpretation and meaning of words, the manner and tone of questions and responses, the likelihood that remarks will be taken out of context, and the prospect that an unexpected or potentially misleading question will be presented during cross-examination. Again, the attorney who has hired the expert will provide the pretrial preparation the expert needs to be effective under cross-examination.

References

Supreme Court of the United States. (1993). *Daubert v. Merrell Dow*, 509 U.S. 579, No. 92-102.

Chapter 9
Application Example 1: Gunshot Acoustics

In the United States, gunshots are among the sounds that may occur in audio forensic recordings. Criminal, terrorist, and accidental actions involving firearms are unfortunately of ongoing concern in many communities, and law enforcement organizations increasingly encounter recorded evidence involving gunfire. Forensic gunshot acoustics is often needed to identify or classify firearms, determine the number and sequential order of multiple shots, and in some cases, estimate the position and orientation of the firearms involved.

Evaluation of recorded gunshot sounds requires a systematic understanding of firearm behavior, as well as the ability to interpret sound wave reflection, absorption, transmission, and diffraction from the ground and other nearby surfaces. The acoustic directionality of the firearm is also an important factor.

In addition to the sound of the gunshot itself, the recorded sounds may include certain mechanical sounds such as loading and ejecting ammunition and cocking the firing mechanism. These relatively subtle sounds may provide important information regarding the type of firearm and other circumstances surrounding a shooting incident.

9.1 Firearm Principles

Ordinary contemporary firearms use ammunition consisting of a *cartridge*. The cartridge, usually made of brass, has a bullet affixed to the open end of a small sealed cylindrical can, called the *case* or the *casing*, containing the propellant (gun powder) charge that typically fills the available space within the cartridge case. Thus, the all-in-one cartridge, sometimes referred to as a *round* of ammunition, provides the proper bullet and propellant load in a single container, as sketched in Fig. 9.1.

The mechanical parts of a firearm have specialized terminology. The specific terms might seem unimportant for audio forensic analysis, but the characteristics and functions of the parts can play an important role in the overall forensic investigation.

R. C. Maher, *Principles of Forensic Audio Analysis*, Modern Acoustics and Signal Processing, https://doi.org/10.1007/978-3-319-99453-6_9

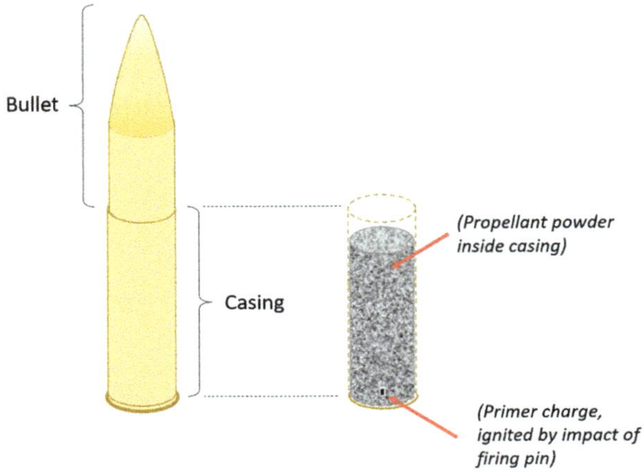

Fig. 9.1 Cartridge containing bullet, casing, and load of propellant

Furthermore, some of the mechanical aspects of the gun can create telltale sounds or can affect the rate of successive shots from a single firearm, and these aspects may be important for forensic acoustics.

The firearm has a hollow cylindrical tube, the *barrel*, through which the bullet will travel when shot. The hollow inside the barrel is known as the *bore*. The cartridge is positioned in a *firing chamber* at the rear end (*breech*) of the barrel. The firing chamber is designed so that the round fits snugly, with the bullet portion of the cartridge protruding into the breech end of the bore. The front end of the barrel, where the bullet emerges, is called the *muzzle*.

The length of the barrel is carefully chosen by the manufacturer on the basis of internal ballistic principles, and the barrel's dimensions affect the level and directionality of the gunshot sound.

The *action* of the firearm refers to the mechanical apparatus that positions the cartridge, allows the spring-loaded *firing pin* to strike the primer cap in the cartridge when the trigger is pulled, and after firing, causes the spent cartridge casing to be removed so that the next round of ammunition can be positioned into the firing chamber.

Centerfire cartridges such as those depicted in Fig. 9.1 contain a primer seated in the head. When the trigger is pulled, the spring-loaded *firing pin* abruptly strikes the cartridge's primer cap, causing it to ignite inside the casing. The primer contains an impact-sensitive mixture which ignites upon the impact of a firearm's firing pin. The burning primer rapidly ignites the propellant surrounding it in the cartridge casing. The hot combustion gases instantly expand, creating extremely high pressure. The pressure release in the casing held snugly in the firing chamber forces the bullet to separate from the casing and accelerate rapidly down the barrel pushed by the hot gas. The bullet emerges from the muzzle with great force and velocity, followed by the high-pressure gas and sooty combustion products that expand out of the muzzle.

Depending on the firearm and the particular load (bullet weight and powder charge), the emerging projectile may or may not be supersonic. If the projectile is supersonic, its passage through the atmosphere will produce a shock wave which can

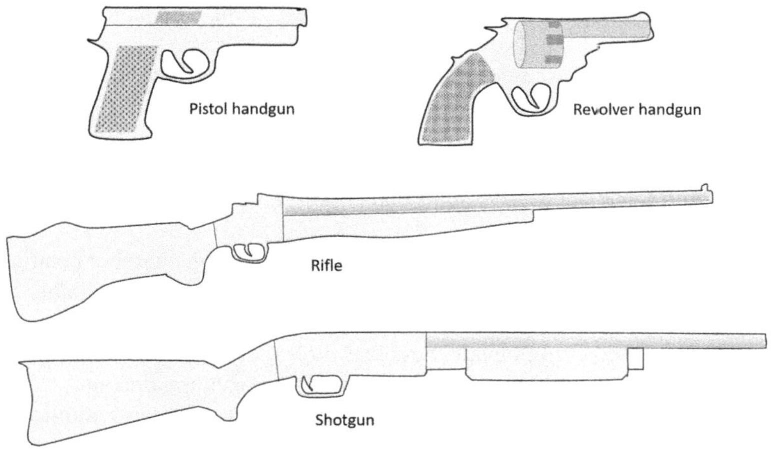

Fig. 9.2 Four basic firearm types

be as loud as the gunshot itself. The supersonic bullet will rapidly lose velocity because of air resistance and will become sub-sonic at some predictable downrange point and velocity. A sub-sonic bullet can still produce sound which may be recorded depending on the quality of the recording device and the proximity of the bullet's passage.

Manufacturers have developed many different cartridge sizes appropriate for the wide variety of firearms in use. The *caliber* of the ammunition is a measurement related to the bullet's diameter. This is specified in inches in the United States. Other countries typically report caliber in millimeters. There can be, an often is, some variability among cartridge designers and manufacturers between the cartridge name and the true caliber of the bullet. For example, the following cartridges all contain bullets whose true diameter is 0.308-inches: .30-30 Winchester, .30-'06 Springfield, .308 Winchester, 7.62 NATO, and .300 Savage. The dimensions of the cartridge cases for each of these examples are different. A firearm will typically be marked by the manufacturer with the cartridge-caliber type, identifying the cartridge that is to be used in the firearm. Furthermore, a particular type of firearm may be fitted with alternative barrels and firing chambers to enable the use of different cartridge types. A particular firearm may be referred to as being *chambered* in a particular caliber, to identify its configuration. The audio forensic examiner need not be knowledgeable about all of these details and can consult with gunsmiths and firearm experts for any relevant details.

The four common standard firearm types are the *pistol handgun*, *revolver handgun*, *rifle*, and *shotgun*. The basic configuration of these firearms is shown in Fig. 9.2.

Handguns are firearms designed to be operated with one hand: the gun has a handgrip, trigger, barrel, and mounted ammunition and is fired by grasping the handle, pointing the gun, and pulling the trigger. Law enforcement officers and target shooters typically grasp the handgun with both hands for better aim and stability.

Many modern handguns and rifles are *repeating* guns, meaning that they have multiple rounds of ammunition stored in an integral compartment, or *magazine*, allowing successive shots without manual reloading. Some magazines hold as many as fifteen cartridges.

Some repeating guns are *semiautomatic*, which means the gun is designed to use the recoil force (or gas pressure) resulting from firing around to eject automatically the spent shell casing and position a new round in the firing chamber without a manual cocking mechanism. Each pull of the trigger for a semiautomatic firearm shoots one bullet, and successive shots can continue with each trigger pull until all of the ammunition is used up.[1]

The term *pistol* refers to a handgun that has a single firing chamber positioned at the back of the handgun's barrel. Examples of commercially available pistols include the Colt M1911 and 1991 Series, Glock 19, and SIG Sauer P320. A pistol typically has a closed firing chamber so that the expanding combustion gases are confined within the gun and the barrel until emerging from the muzzle.

The term *revolver* refers to a handgun with a *cylinder* containing multiple firing chambers. To prepare the revolver, a cartridge is pre-loaded into each chamber in the cylinder. The cylinder rotates after each shot so that an unspent cartridge is positioned behind the barrel, ready for the next shot. Most, but not all revolvers possess six chambers in their cylinders. Examples of commercially available revolvers include the Ruger SP101 and the Smith & Wesson Model 629. A revolver, with its multiple chambers that rotate into position for firing, has a small gap between the firing chamber and the gun barrel at the point where the rotating cylinder and the back of the gun barrel meet. Some of the hot combustion gases may leak out from the cylinder-barrel gap when a revolver is fired.

A *rifle* has a relatively long barrel, typically more than 40 cm in length, affixed to a frame *stock* so that the gun is fired from the shoulder while being held with both hands. One hand supports the fore end of the gun and the other grasps the stock just behind the trigger. A rifle fires a single bullet at a time. The term "rifle" comes from a simplification of the term "rifled musket," referring to the helical grooves, or *rifling*, cut into the bore of the barrel. The rifling imparts a spin on the bullet that helps stabilize its trajectory, much like what we see in the "spiral pass" of a well-thrown football.

The type of *action* employed in rifles will fall into one of the following categories:

- *Bolt action*— a manual bolt handle and bolt at the rear of the firing chamber which allows the user to chamber and secure a live cartridge in the firing chamber and to manually extract the fired cartridge from the chamber. Bolt-action rifles may be single shot or may contain a number of live cartridges in a magazine located below the bolt.
- *Lever action*— a manually-operated lever just below and behind the trigger which is rotated downward to extract and eject a fired cartridge and to load a live cartridge into the firing chamber with an upward, closing movement of the lever.

[1] Note that unlike a *semi*-automatic firearm, a gun equipped with a *fully-automatic* action, is by design and definition, a machine gun. Such firearms continue to fire at a very high cyclic rate as long as the trigger is held rearward, or until it is released. In the United States, fully-automatic firearms are highly regulated and restricted to the military, law enforcement and certain specially-licensed citizens.

Lever-action rifles are usually repeaters and contain an ammunition supply in a tubular magazine under the barrel or in an area below the breech block.

- *Pump/slide action*—a manually-operated sliding mechanism located forward of the action and below the barrel. Retracting the pump handle extracts and ejects a spent cartridge; the forward movement loads a live cartridge into the firing chamber. Slide-action rifles are repeaters and contain an ammunition supply in a tubular magazine under the barrel.
- *Break action*— the barrel of this type of rifle is hinged at the receiver in such a way that it can be manually opened allowing a live cartridge to be inserted in the chamber. The closing of the mechanism readies the gun for firing. It must be re-opened to remove the spent cartridge from the chamber. Such rifles are single shot guns.
- *Semiautomatic action*—These rifles typically use some of the high-pressure powder gases to unlock and open the action, then extract and eject the fired cartridge case. A spring mechanism returns the breech block (bolt) forward stripping a live cartridge from the magazine and loading it in the firing chamber. Examples of such firearms are the AK47 and AR15. As with semi-automatic pistols, each shot with a semi-automatic rifle requires a pull of the trigger.

Like a rifle, a *shotgun* also has a long barrel and stock so that it is fired from the shoulder. Unlike rifles, these guns possess smooth (no rifling) bores of large diameter and use ammunition of relatively complex and varied design. The typical, modern shotgun cartridge used for sporting purposes, often called a *shotgun shell*, has a plastic body affixed to a brass-plated steel head, which also contains a centrally-located primer. The contents consist of a large number of spherical lead or steel *pellets*, or *shot*, ranging in diameter from about 0.08-in. (2mm) to 0.13-in. (3.3mm). A sketch of a typical shotgun cartridge is shown in Fig. 9.3.

There are also much larger diameter pellets, called buckshot which have diameters on the order of 0.24-in. (6mm) to 0.33-in. (8mm). These pellets may be nested

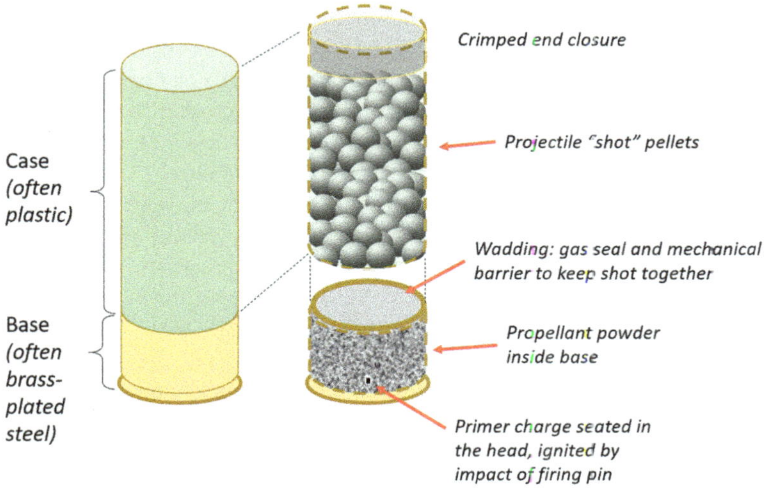

Fig. 9.3 A shotgun cartridge containing multiple pellets or "shot"

in a plastic *shotcup* or reside on top of some thick, cardboard wadding. The powder charge is below the shotcup or wadding. The cup or wadding seals off the burning powder gases from the shot charge. As the name suggests, buckshot was designed for hunting deer. Buckshot cartridges also have a law enforcement application.

Another type of shotgun cartridge contains a single projectile called a *solid slug*, or simply a *slug*. Finally, there are special-purpose shotgun shells which contain rubber pellets, chemical munitions, or so-called *bean bags*. These are used primarily by law enforcement for less lethal purposes (animal control, dispersing rioters, or subduing a dangerously agitated or aggressive subject).

9.2 Firearm Acoustics

Audio forensic analysis of gunshots usually involves the obvious and loud "bang" of the gun, but some investigations make use of the telltale sounds of the firearm's action, spent cartridge ejection, and positioning of new ammunition. These characteristic sounds may be of interest for forensic study if the microphone is located sufficiently close to the firearm to pick up this subtle sonic information. Audio forensic analysis of recorded gunshots can help verify eyewitness (and "ear" witness) accounts and aid in crime scene reconstruction (Weissler and Kobal 1974; Brustad and Freytag 2005). The gunshot recording may also be analyzed to identify acoustical reflections and other sonic effects of the gunshot process (Koenig et al. 1998).

9.2.1 *Muzzle Blast*

A conventional firearm uses a confined combustion of gunpowder to propel the bullet out of the gun barrel. The hot, expanding gases rapidly pressurize the chamber behind the bullet, abruptly forcing a supersonic jet of gas from the muzzle. The sound of the gunshot, the *muzzle blast*, is emitted from the end of the muzzle in all directions, but the majority of the acoustic energy is expelled in the direction the gun barrel is pointing. The muzzle blast causes an acoustic shock wave and a brief, chaotic sound lasting only a few milliseconds (Beck et al. 2011).

The muzzle blast sound propagates through the air at the speed of sound (e.g., 343 m/s at 20 °C) and interacts with the surrounding ground surface, obstacles, temperature and wind gradients in the air, spherical spreading, and atmospheric absorption. If a recording microphone is located close to the firearm, the direct sound of the muzzle blast is the primary acoustical signal. On the other hand, if the microphone is located at a greater distance from the firearm, the direct sound path may be obscured, and the received signal will exhibit propagation effects, multipath reflections, and reverberation (Maher 2006, 2007; Maher and Shaw 2008).

In the United States, various regulations prevent the widespread use of *acoustic suppressors*, sometimes called "silencers," with firearms. Suppressors are designed to reduce the audible report (and often the visible explosive flash) of the muzzle blast to reduce the likelihood of detection and/or to prevent hearing damage. Although Hollywood movies often depict a "silencer" as being incredibly effective at eliminating the muzzle blast sound, in reality a suppressor may reduce the peak sound intensity by some amount, but the gunshot is still clearly audible and noticeably loud.

9.2.2 Mechanical Action

For some firearms the sound of the mechanical action may be detectable if the microphone is close enough to the gun. This includes the sound of the trigger and firing pin mechanism, the ejection of spent cartridges, and the positioning of new ammunition by the gun's semiautomatic or manual loading system. Because these subtle, telltale sounds are generally much quieter than the muzzle blast, their presence may only be detectable from personal surveillance recordings or possibly recorded telephone conversations if the gunfire is nearby.

9.2.3 Supersonic Projectile

If the bullet emerging from the barrel is moving at a speed greater than the speed of sound in the air, it is a *supersonic* projectile. A supersonic projectile moves too quickly for the surrounding air particles to react with the normal relationship between particle pressure and particle velocity described by linear acoustics. Instead, the supersonic projectile launches an acoustic *shock wave* in the air as the projectile travels downrange from the firearm. The ballistic shock wave expands like a cone trailing the bullet. The shock wave front propagates at the speed of sound, so the angular direction of the shock wave depends upon the bullet's speed with respect to the speed of sound (Sadler et al. 1998).

If the speed of sound in the air is c m/s and the projectile is traveling at speed V m/s, the projectile's *Mach Number* is given by $M = V/c$, a dimensionless quantity sometimes designated in "mach." A supersonic projectile will have a Mach Number greater than 1, because $V > c$. The ballistic shock wave trailing the supersonic bullet has an inner angle $\theta_M = \arcsin\left(\frac{1}{M}\right)$, which depends upon the Mach Number. A projectile with speed just barely above the speed of sound, i.e., a Mach Number just over 1 mach, will have a broad shock wave cone ($\theta_M \rightarrow 90^\circ$), while a high-velocity rifle bullet with, say, $M = 3.5$ mach, will have a narrow shock wave cone ($\theta_M \rightarrow 16.6^\circ$). See Fig. 9.4 for a pair of shock wave sketches, one for a bullet traveling 1.8 mach and the other traveling 3.5 mach.

As noted previously, the speed of sound (c) in air increases with increasing temperature:

$$c = c_0\sqrt{1 + \frac{T}{273}}$$

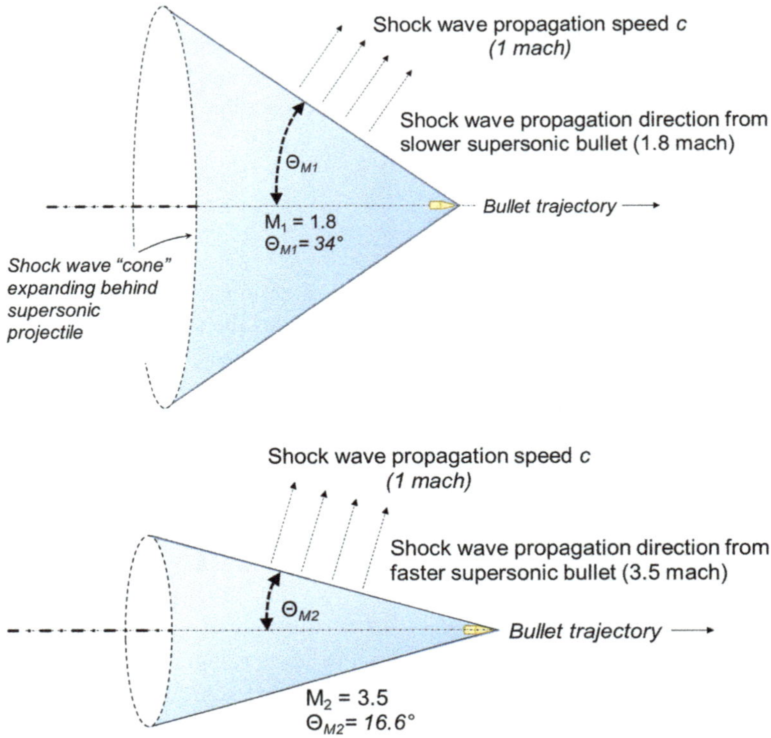

Fig. 9.4 Ballistic shock wave characteristics for supersonic bullets. Faster supersonic projectiles have narrower shock wave cone angles

where T is the air temperature in degrees Celsius and $c_0 = 331$ m/s is the speed of sound at 0 °C. For each degree Celsius increase in temperature, the speed of sound increases by approximately 0.61 m/s.

If the bullet is traveling substantially faster than the speed of sound, the Mach Angle is small and the shock wave propagates *nearly perpendicularly* to the bullet's trajectory. A bullet traveling only slightly faster than the speed of sound has a Mach Angle approaching 90°, meaning that the shock wave is propagating *nearly parallel* to the bullet's path. Moreover, as the bullet slows along its path due to friction with the air, the bullet's Mach Number decreases, and the corresponding Mach Angle widens downrange.

While the muzzle blast sound propagates away from the firearm at the speed of sound, a supersonic bullet is traveling faster than sound, so the bullet zooms downrange outpacing the muzzle blast sound. If a microphone is located downrange, the first pressure disturbance it may detect will be due to the supersonic bullet's shock wave trailing the bullet, followed by the muzzle blast sound.

9.2.4 *Surface Vibration*

In some circumstances, the impulsive sound of a gunshot may cause vibration of the ground or other nearby surfaces. Sound vibrations launched in solid materials such as surface rock and soil may travel at least five times faster than the speed of sound in air. If the surface vibration results in a detectable signal at a microphone mounted to the surface some distance away, it may be feasible to compare the time of arrival of the vibratory signal and the airborne signal to resolve uncertainty about the firearm's location.

In summary, the primary acoustical evidence available from a gunshot can include the muzzle blast, the projectile shock wave for supersonic bullets, and possibly the sound of the firearm's mechanical action and ground vibration, if the microphone is sufficiently close to the gun.

9.3 Example Demonstration Gunshot Recordings

Gunshot recording and analysis is a specialized field due to the intense and impulsive nature of the muzzle blast and the projectile's shock wave, if present. The peak sound pressure levels near the firearm can exceed 150 dB re 20 μPa. The high peak pressures associated with the gunshot sounds can cause clipping in the microphone and the preamplifier input stage, and the extremely rapid rise times are usually sufficiently distorted by the recording system to make quantitative observation difficult. This is particularly true for audio forensic recordings obtained via telephone.

Although the gunshot sounds used in Hollywood movies and video game soundtracks are usually many hundreds of milliseconds in duration, the actual duration of a firearm's muzzle blast is typically only 1–3 ms, and in the case of a supersonic bullet, the ballistic shock wave over- and under-pressure signature is just a few hundred microseconds in duration. Indeed, from a forensics standpoint, the "sound effects library" gunshot recordings tell more about the acoustical impulse response of the area surrounding the recording position than they do about the firearm itself, because such recordings deliberately contain an artificially high level of echoes and reverberation to enhance the emotional impact.

In fact, earwitnesses who hear true gunfire often remark that the sounds seemed like mere "pops" or "firecrackers" rather than gunshots, at least in comparison with their media-influenced expectations. Even if reverberation is not added deliberately, gunshot recordings obtained in acoustically reflective areas, such as indoors or outdoors in an urban area, may contain a mixture of overlapping shots and echoes that can complicate the analysis process (Beck et al. 2011).

Thus, to begin our thorough consideration of audio forensic analysis of gunshot sounds, we start with example recordings made under controlled conditions with specialized professional recording equipment. The recording levels, sampling rates, and microphone positions were carefully controlled to avoid clipping and acoustic reflections.

As will be seen later, the observed details in these "laboratory" recordings are often obscured in real forensic recordings obtained under uncontrolled circumstances with common speech band mobile devices. Nevertheless, it is important to understand the true characteristics of the underlying gunshot signal so that the effects and limitations of the real acoustic environment and recording systems become clear.

9.3.1 *Rifle Shot with Supersonic Projectile*

As a first example, a demonstration audio recording of a rifle shot involving supersonic ammunition is shown in Fig. 9.5. The rifle was fired at shoulder height, and the microphone was located approximately 7 m away at 45° azimuth. The recording was made with a 48 kHz sampling rate, 16-bit resolution, and a professional omnidirectional recording microphone (DPA 4003 with high-voltage preamplifier HMA 5000).

Even in this relatively simple geometry with a single shot and a single prominent reflection from the ground, the resulting waveform is more complicated than might be expected (Maher 2007). The prominent features and a plan view of the test orientation are shown in Fig. 9.6.

The first portion of the sound is the acoustic shock wave from the supersonic bullet. The shock wave has a distinctive and extremely abrupt onset and offset, creating what is referred to as an "N" wave. At low acoustic amplitudes where the ratio

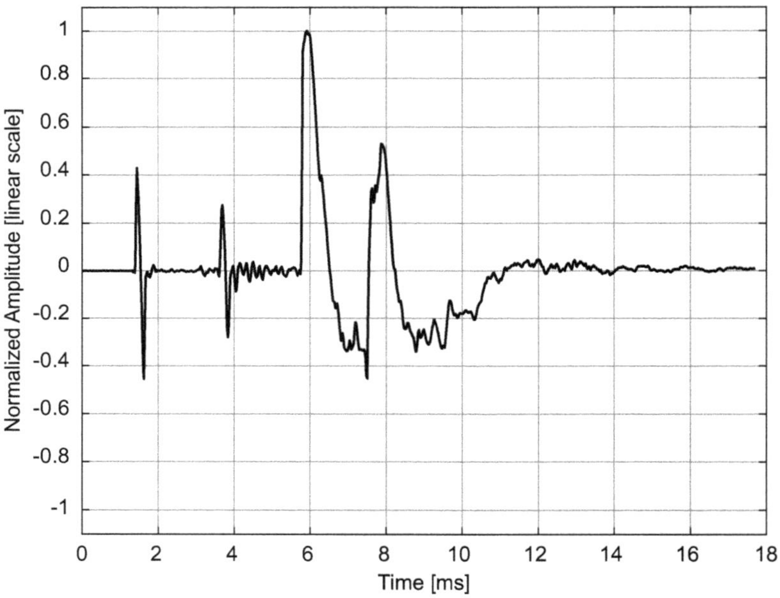

Fig. 9.5 Gunshot recording, 0.308 Winchester rifle fired horizontally over the firm ground of a shooting range, speed $M = 2.54$ mach, oblique trajectory past the microphone

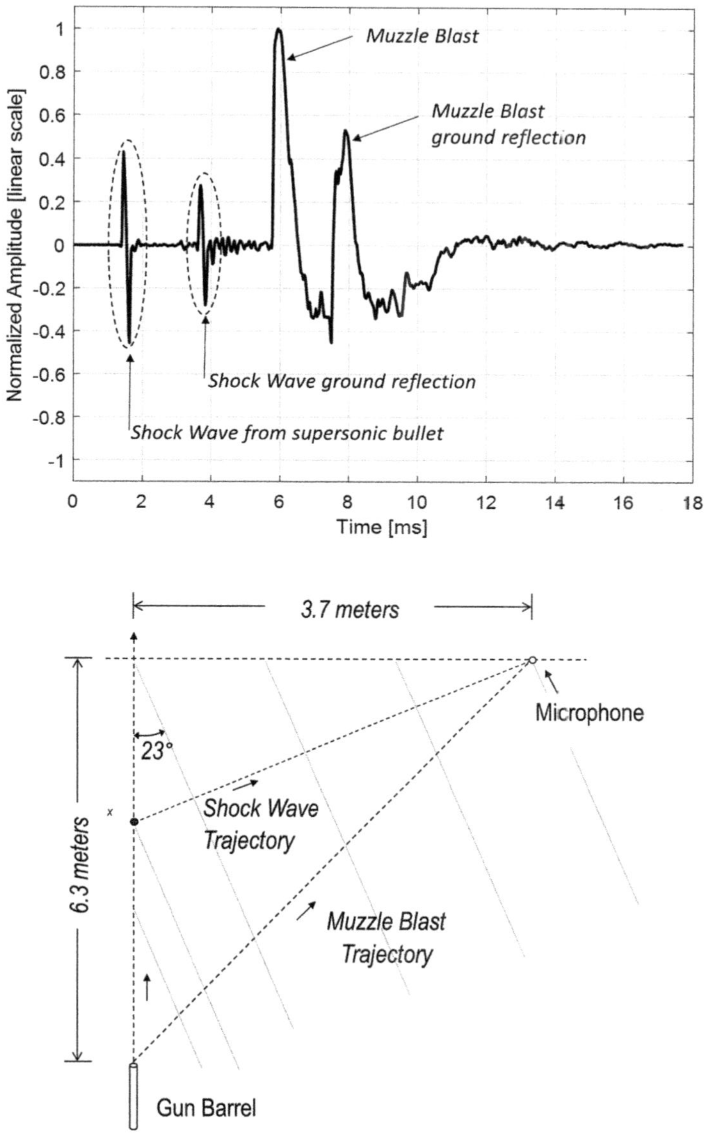

Fig. 9.6 Annotated waveform description (ref. Fig. 9.5) and shot orientation (plan view, not to scale)

of pressure to particle velocity is a linear relationship, the speed of sound is essentially independent of the waveform's amplitude. But with a strong shock wave, high-amplitude nonlinearity of the air results in a difference in wave front speed between the high-amplitude portion (faster speed) and the low-amplitude portion (lower speed). This causes the peaks of the shock wave disturbance to propagate

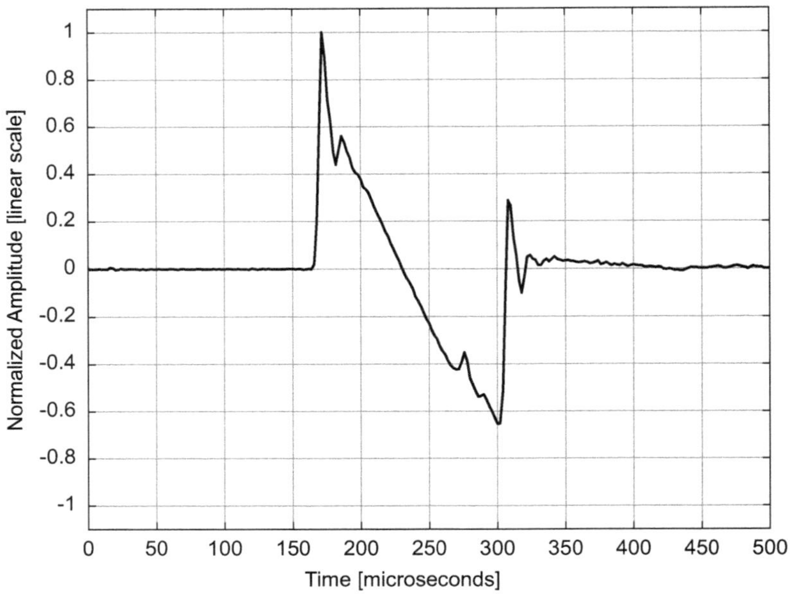

Fig. 9.7 Acoustic shock wave disturbance recorded from a supersonic 0.308 Winchester bullet: the "N" wave

faster than the low-amplitude portions, resulting in the characteristic "N" shape of the pressure disturbance. A time-expanded example of a recording of a ballistic shock wave is shown in Fig. 9.7. The waveform was recorded using a specialized microphone system and 500 kHz sampling rate (G.R.A.S. type 46DP 1/8" microphone set with type 12AA power module).

After the arrival of the ballistic shock wave, the next event in the recording is the arrival of the shock wave reflected from the ground. The reflection is somewhat lower in amplitude than the direct sound of the shock wave due to the slightly greater distance traveled (down to the ground and back up to the microphone) and the energy absorbed by the ground.

The arrival of the muzzle blast sound occurs after the ballistic shock wave because the supersonic projectile projects the shock wave downrange, closer to the microphone. The direct sound of the muzzle blast is followed by the muzzle blast sound reflected from the ground. In this geometry the direct muzzle blast sound and the reflected sound partially overlap in time.

It is very important to notice that the relationship between the shock wave, muzzle blast, and ground reflections depend upon the orientation and distance between the firearm and the microphone, as well as the speed of the projectile, the speed of sound, the reflection and absorption of the ground surface, and a number of other factors. Consider the recording of the same 0.308 Winchester ammunition fired in a different orientation with respect to the microphone, as shown in Fig. 9.8.

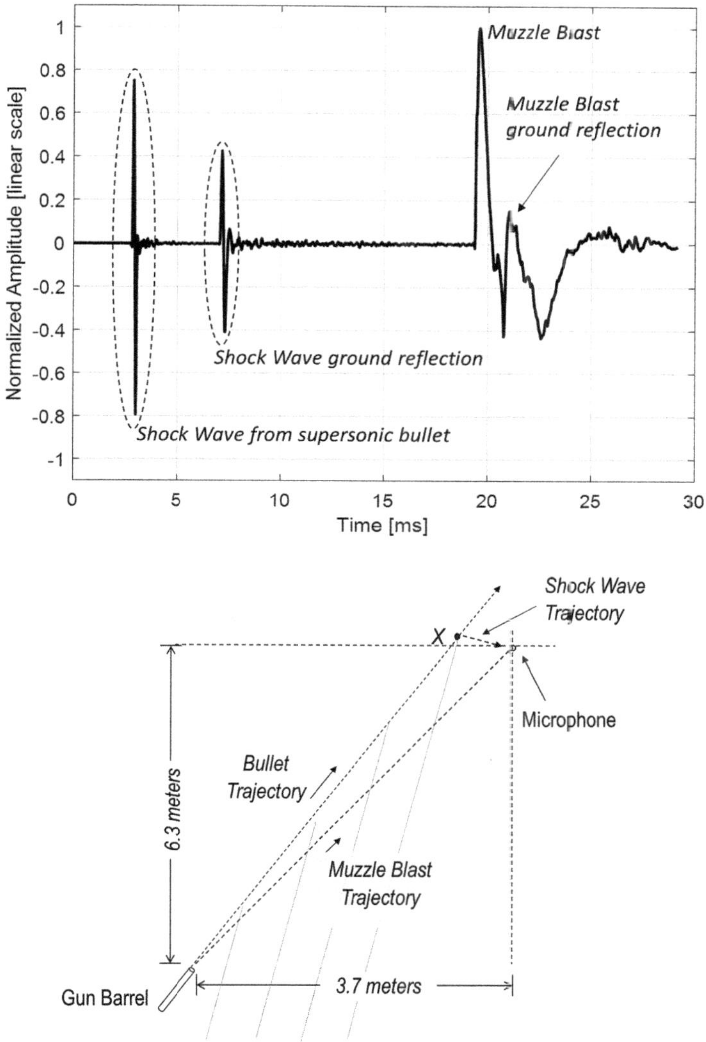

Fig. 9.8 Annotated waveform description and altered shot orientation (plan view, not to scale). Compare to Fig. 9.6 (and note different time scale)

Even with the same ammunition and shooting position, the details of the waveforms differ between Figs. 9.6 and 9.8. In particular, the timing between the shock wave arrival and the muzzle blast arrival is greater in the latter case in which the bullet's trajectory comes close to the microphone, because the supersonic bullet projects the shock wave toward the vicinity of the microphone faster than the speed of sound. Moreover, the details of the direct sound and reflected sound also vary with the orientation of the firearm with respect to the microphone, and the differences are even greater if additional acoustical reflections and reverberation are present, if multiple shots are overlapping, or if the recording microphone does not have direct line of sight with the firearm.

A common misconception is that the muzzle blast of a gunshot can be considered an omnidirectional impulse point source of sound. In fact, the muzzle blast is quite directional, with a substantially higher sound pressure amplitude in the direction the muzzle is pointing compared to the level at angles farther off-axis (Maher 2010). From the standpoint of audio forensic analysis, this fact means that interpreting a forensic recording containing gunfire must also address the signal characteristics due to the firearm's spatial orientation.

In Fig. 9.9, results of a special directional recording technique are shown for a 0.308 rifle shot (Routh and Maher 2016). The recording was made outdoors with an elevated shooting position and an elevated set of microphones positioned in a semi-circle around the firearm (see Figs. 9.10 and 9.11), so that the shot acoustics can be recorded before the first acoustic reflection (from the ground) arrived at the microphone position. The gun is fired by a marksman holding the firearm in the conven-

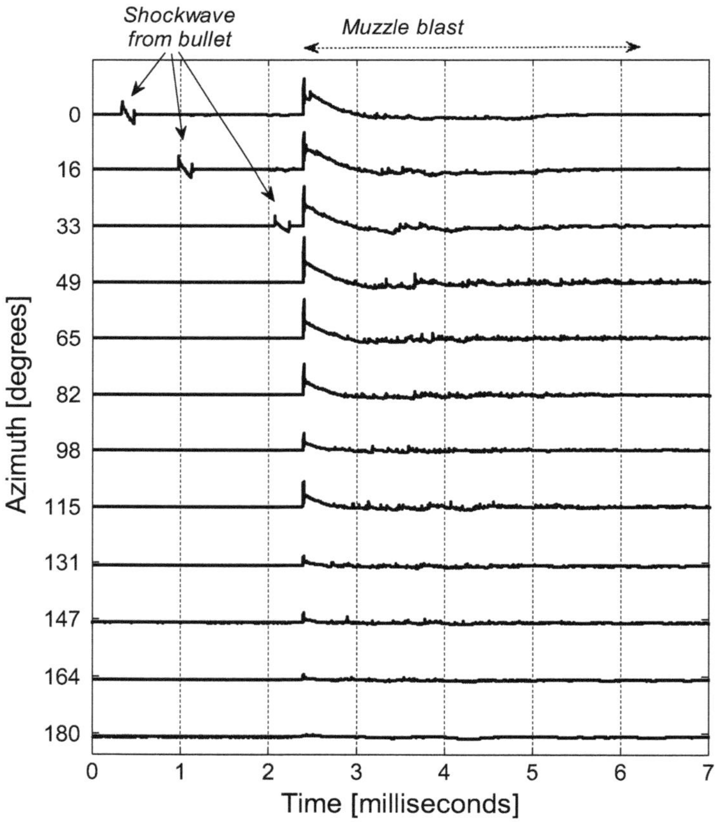

Fig. 9.9 Example quasi-anechoic directional recording of a 0.308 rifle shot. Each trace depicts the sound recorded at the corresponding azimuth with respect to the direction the barrel was aimed (zero azimuth)

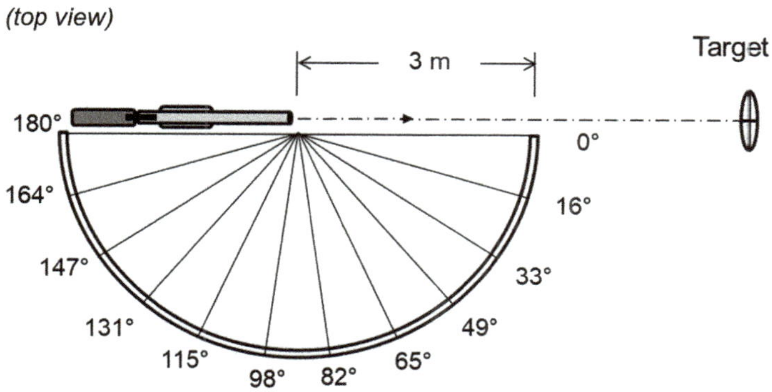

Fig. 9.10 Plan view diagram (not to scale) depicting multi-microphone recording orientation. The typical position of the marksman holding and firing the gun in the conventional manner is not shown (Routh and Maher 2016)

Fig. 9.11 Research recording setup used to obtain quasi-anechoic gunshot data for 12 azimuths from 0 to 180°. The marksman's head, arms, and torso cause natural reflections and diffraction, especially for the microphones located at the azimuths 90–180°. The firearm and the microphones are elevated so that the full duration of the muzzle blast can be recorded prior to the arrival of the first major reflection from the ground (Routh and Maher 2016)

tional manner, so the sound waves from the gunshot sound can reflect from and diffract around the shooter's arms and body. The quasi-anechoic recording process provides a unique way to visualize the firearm's acoustical behavior.

The directional recordings of Fig. 9.9 express several key observations for forensic audio analysis of gunshots.

First, as noted previously, the supersonic bullet's ballistic shock wave propagates outward and forward as a cone trailing the bullet. As the bullet leaves the muzzle

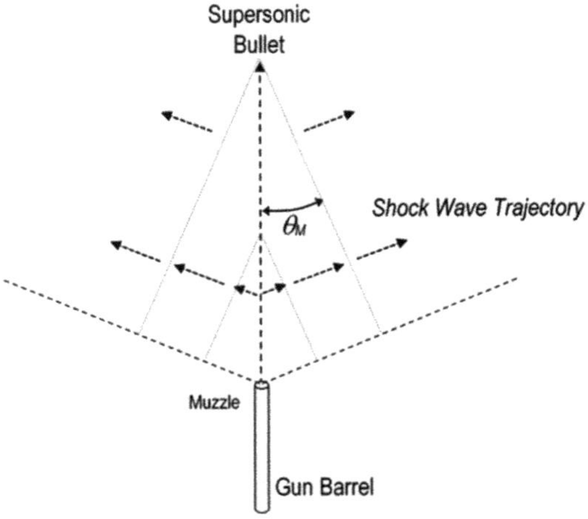

Fig. 9.12 Depiction of supersonic shock wave propagation direction and orientation

and starts triggering the shock wave, the angle of the shock wave front does not extend to the sides or to the rear of the gun, so in this example the shock wave is only detected out to just beyond ~30° in azimuth (see Fig. 9.12). Thus, the presence of a recorded shock wave will indicate a supersonic projectile, but the absence of a shock wave signature in a recording does not necessarily imply a subsonic bullet if the microphone is at an azimuth greater than the shock wave trajectory.

Second, the direct sound of the muzzle blast is very short in duration: just 3–4 ms even for a substantial rifle shot. As noted previously, inexperienced witnesses near a shooting scene often remark that the sounds they heard didn't sound like gunshots but more like the "pop, pop, pop" of firecrackers. The expectation of many earwitnesses is skewed by the gunshot sound effects typically used in Hollywood movies, television, and computer games, which are highly reverberant and last 1–2 s, not the 3–4 ms of a real muzzle blast.

Third, it is important to note the directionality of the firearm's muzzle blast sound. In this case, for the 0.308 rifle shot, the sound level toward the rear is approximately 20 dB lower than the sound level in the direction the rifle is pointing, as shown in Fig. 9.13.

9.3.2 Pistol Shot with Subsonic Projectile

For a second example, a quasi-anechoic audio recording of a handgun pistol shot at approximately zero azimuth (gun pointed at microphone) is shown in Fig. 9.14. The majority of the sound energy in the muzzle blast lasts only about 1 ms. Because the bullet is traveling at a speed less than the speed of sound, no ballistic shock wave is present.

The same pistol shot observed off-axis at 98° azimuth is shown in Fig. 9.15. Note that the signal details differ as a function of azimuth, even for the same shot from the

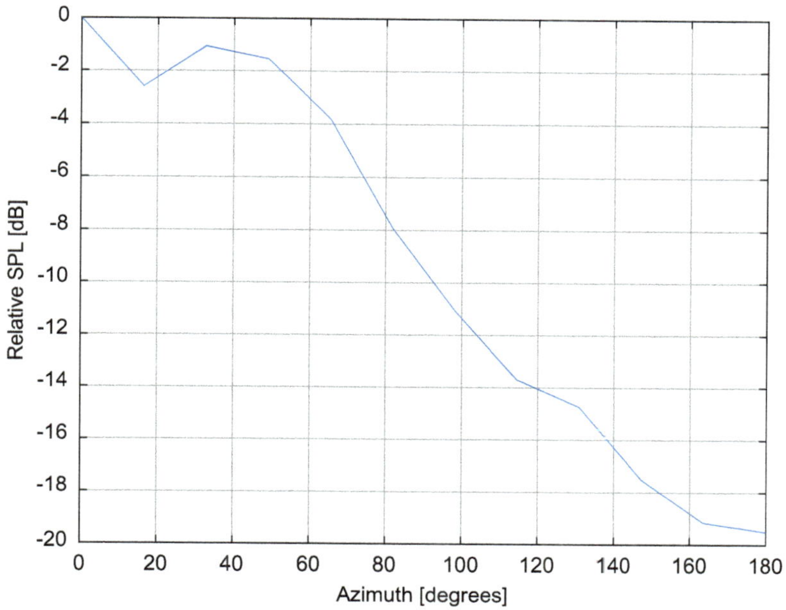

Fig. 9.13 Measured dependence of sound pressure level (muzzle blast portion) vs. azimuth for a 0.308 rifle shot

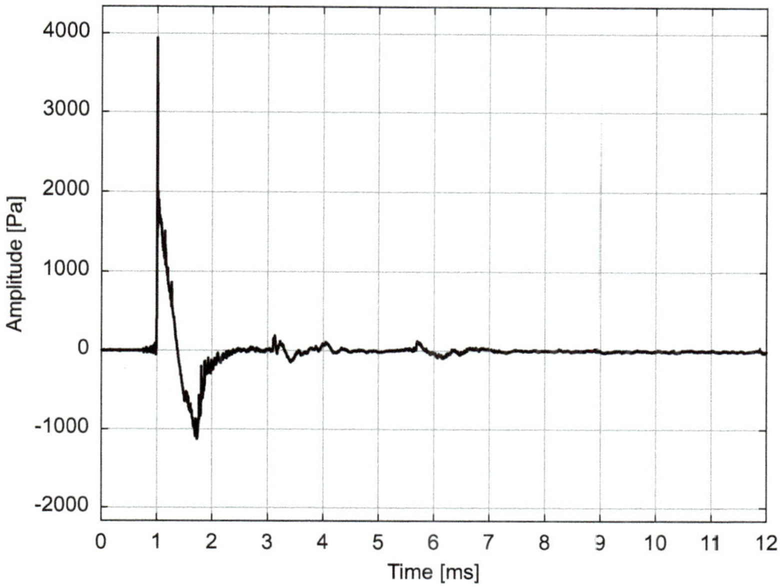

Fig. 9.14 Pistol (Glock 19, 9 × 19 mm) muzzle blast recorded at 3 m distance and on-axis at approximately zero azimuth

same firearm. This fact is important to understand, because some gunshot classification software packages are trained using a limited example database that does not represent the azimuth dependence of the gunshot waveforms. In fact, it is important for audio forensic examiners to recognize that the difference in level and waveform details between on-axis and off-axis recordings of the *same* firearm are often significantly greater than the difference between two *different* firearm types at the same azimuth. This can have an important effect upon properly deducing the firearm type from a recording, especially if the orientation of the firearm with respect to the microphone is not known from some other source of information.

9.3.3 Revolver Shot with Subsonic Projectile

For a third example of carefully controlled demonstration, a quasi-anechoic audio recording of a handgun revolver shot at approximately zero azimuth (gun pointed at microphone) is shown in Fig. 9.16, and the same shot recorded at azimuth 98° is shown in Fig. 9.17.

The on-axis waveform of the pistol (Fig. 9.14) and the revolver (Fig. 9.16) are quite similar, but the off-axis characteristic of the revolver (Fig. 9.17) shows a notable difference. From the side of the firearm, the acoustical signal includes an impulse due to the sound coming from the gap between the revolver's cylinder and

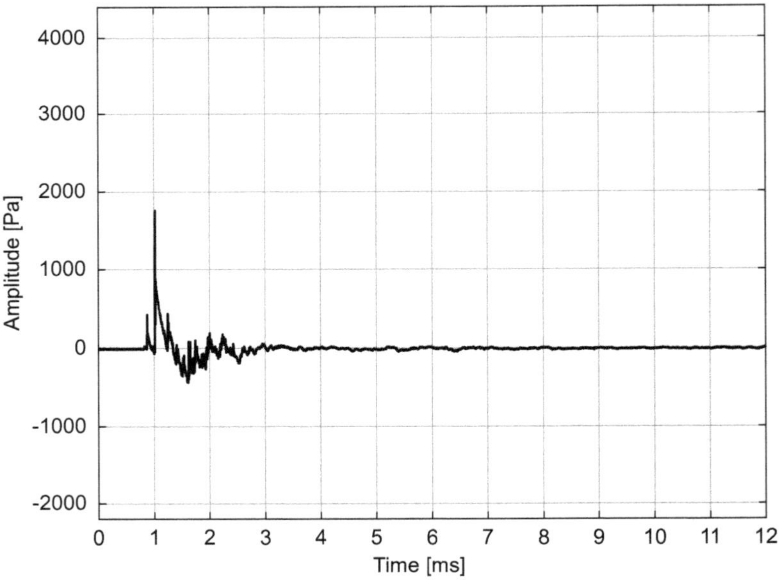

Fig. 9.15 Pistol (Glock 19, 9 × 19 mm) muzzle blast recorded at 3 m distance and off-axis at approximately 98° azimuth

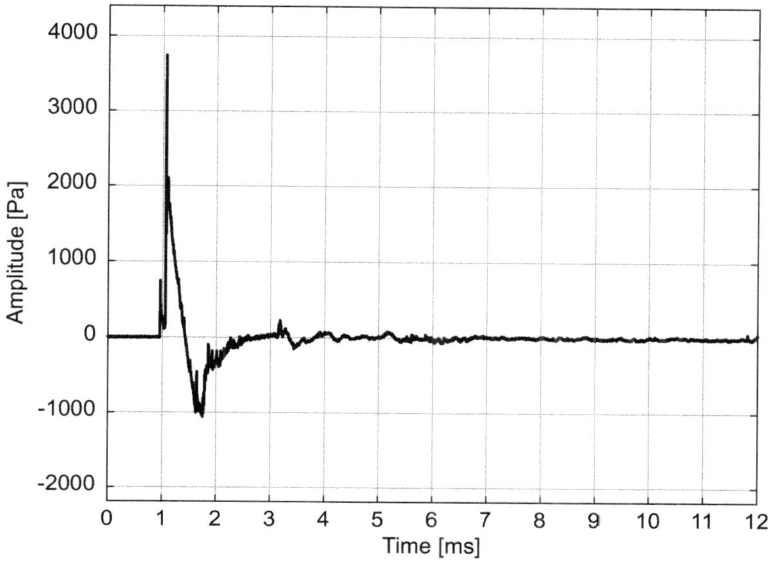

Fig. 9.16 Revolver (Ruger SP101, 38 special) muzzle blast recorded at 3 m distance and on-axis at approximately zero azimuth

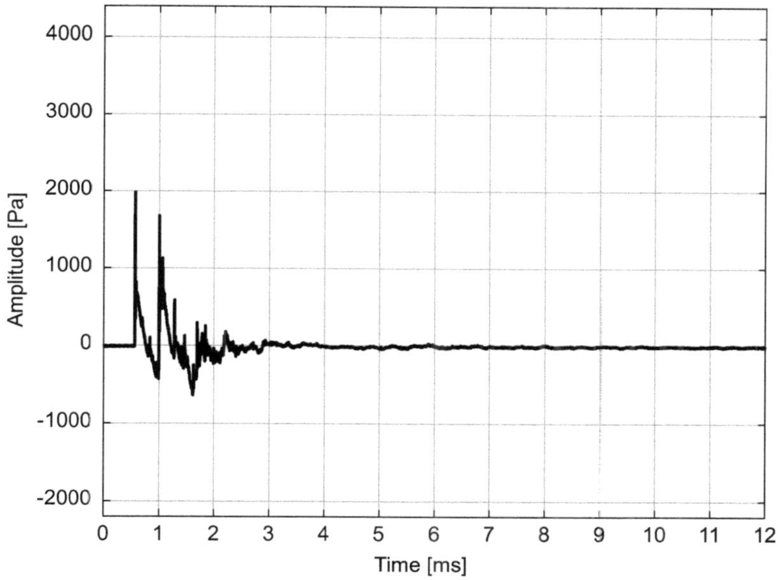

Fig. 9.17 Revolver (Ruger SP101, 38 special) muzzle blast recorded at 3 m distance and off-axis at approximately 98° azimuth. Note the presence of two signal peaks: one due to sound emerging from the cylinder gap and the other from the muzzle

the breech end of the barrel, along with the sound of the muzzle blast emanating from the muzzle of the barrel (Maher and Routh 2017). The time difference between these two sounds depends upon the length of the barrel and the time required for the bullet to accelerate upon firing and emerge from the barrel.

9.4 Example Forensic Gunshot Recordings

While the previous example recordings demonstrate gunshot properties obtained under carefully controlled conditions with specialized microphones and high sampling rates, the recordings most commonly encountered by audio forensic examiners are clipped, distorted, reverberant, and difficult to decipher.

9.4.1 Example Forensic Recording 1: Gunshots, Taser, and Speech

Figure 9.18 shows the waveform and spectrogram of a 14-s audio excerpt taken from a forensic recording. The labels in the figure were entered manually.

This particular recording came from the soundtrack of a video produced by an audiovisual attachment for a *TASER*® brand Conducted Electrical Weapon (CEW), or what many in law enforcement refer to as an Electronic Control Weapon (ECW). An ECW, such as a Taser, is a weapon designed to be less lethal than a conventional firearm for use by law enforcement officers. The device is held and aimed like a handgun, but instead of firing bullet cartridges propelled by gunpowder, the Taser uses compressed air cartridges to propel a pair of darts toward the targeted individual. Each dart trails a thin conductive wire. Upon the darts striking and embedding in the target, the device generates a rapid sequence of electrical impulses between the darts through the conductive wires. The ECW's electrical pulses are designed to disrupt the target individual's neuromuscular control, causing the person to stop fighting, fleeing, or resisting the officer's commands. The pulses occur approximately every 50 ms and continue for 5 s or a shorter duration if the device's batteries are partially discharged. Pulse generation can be continued for more than 5 s if the officer continues to depress the trigger.

When the Taser device is deployed, its electrical circuitry may cause an audible, rapidly pulsating sound if the darts did not fully engage the target. In an audio recording, the Taser device's 50 ms electrical pulse period may appear as a rapid train of impulses in the recording. These distinctive pulses can be seen in the upper left corner of Fig. 9.18, as indicated in the figure.

The six gunshots in this example recording came from one or two 9 mm handguns. One was known to be a Glock 19 pistol, and the investigators suspect there may have been another 9 mm handgun discharged at the scene. Unlike the pristine, echo-free reference recording of a pistol shot that lasts 2–3 ms (see the Glock 19

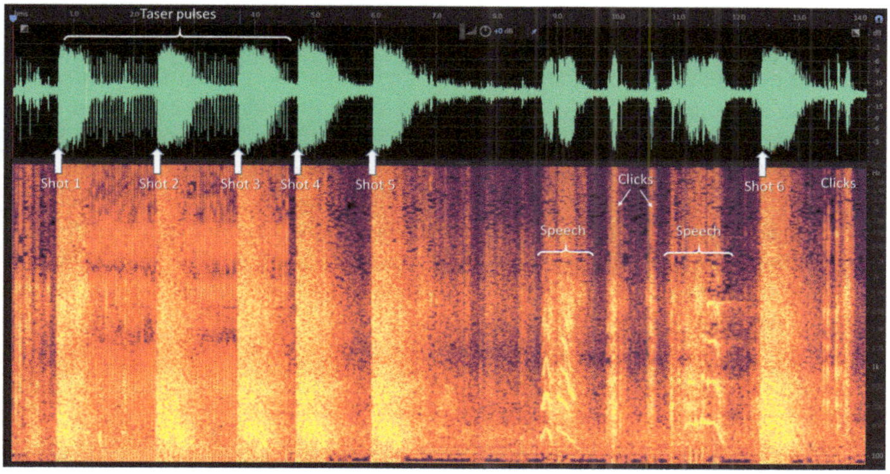

Fig. 9.18 Example forensic audio recording with gunshots, Taser pulses, and speech utterances (overall duration 14 s, frequency range 0–4 kHz, linear scale)

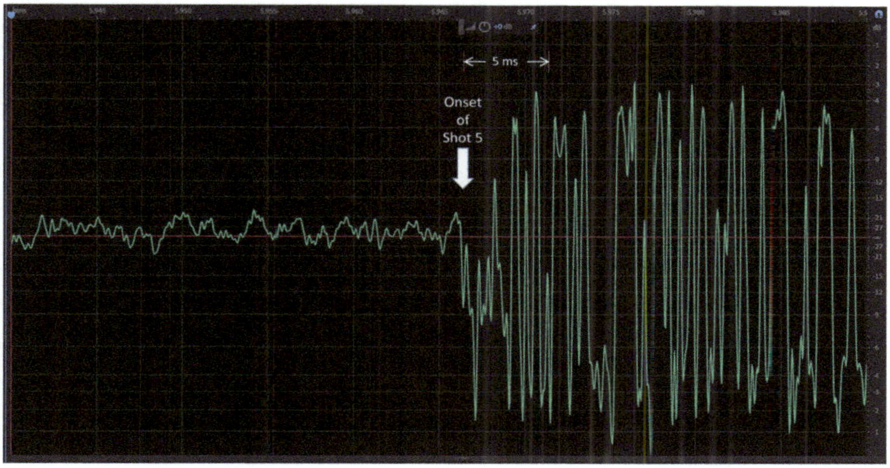

Fig. 9.19 A 50 ms portion of the forensic recording shown in Fig. 9.18, centered on "Shot 5 (overall duration 50 ms)"

example in Fig. 9.14), the acoustical signature of the shots in this evidentiary recording occupies over 700 ms due to numerous overlapping sound reflections and reverberation.

A small portion of the recorded waveform corresponding to the sound labeled "Shot 5" in Fig. 9.18 is shown in Fig. 9.19. To reiterate, it is essential to keep in mind that the acoustic energy from the Glock 19 pistol's muzzle blast only lasts 2–3 ms (Fig. 9.14), so the reverberant signal lasting hundreds of milliseconds in this forensic recording is dominated by reflected and reverberated sound recorded at the scene.

Based on the video that accompanied the audio recording, a detective reported that the officer prepared to use the Taser device by moving the safety switch to the ARMED position, because that is the action necessary to initiate the audiovisual recording. According to the detective who reviewed the video, the shadowy scene initially shows that the Taser device is pointed toward a fleeing suspect and then pointed downward without taking a shot. The Taser was subsequently dropped to the ground approximately 800 ms before the first gunshot. The detective believes that the Taser *might* have fired due to striking the ground, not by the officer deliberately pulling its trigger. In any case, for the audio recording time span shown in Fig. 9.18, the corresponding video depicts only unmoving ground debris, and the detective believes that the dropped Taser device was lying still on the ground in a fixed position.

Hypothetically, the detective could ask several questions for the audio forensic examiner, such as:

1. Are the gunshot sounds all attributable to a single firearm?
2. Is the Taser deployment before or after the first gunshot?
3. Were the speech sounds uttered by a man or by a woman?

Assuming there is no dispute about the authenticity of the recording, the audio forensic examiner's process would be to conduct a quick aural evaluation, followed by critical listening, waveform analysis, and spectral analysis.

Question 1: A Single Firearm?

Question 1 requires several determinations. From *critical listening*, the six gunshots are found to be distinct and not overlapped. *Subjectively*, they sound similar in terms of the perceived loudness, timbral quality, and reverberation tail duration. However, listening to pairs of the shots in succession, e.g., Shot 1 and Shot 2, Shot 1 and Shot 3, Shot 1 and Shot 4, etc., gives a *subjective* impression that there is a noticeable difference between the first two shots and the last four shots. This critical listening could indicate that different firearms were used or that there was some difference in the position and/or orientation of a single firearm with respect to the recording device (the Taser), which was known to be stationary on the ground. While the Taser was not moving, the officer believed to be shooting the Glock 19 likely *was* moving and turning, which could account for the audible differences.

From *waveform analysis*, the timing and energy envelope of each shot can be compared. If the Glock 19 was the only firearm discharged, a question could be raised about how rapidly it is possible to fire successive shots from that gun. From the waveform analysis, the timing of the six shots indicates a minimum inter-shot time gap of 0.973 s. This information from the audio forensic analysis could be used by a firearms expert to predict whether or not the gun could be fired that quickly. Studies have shown that a semiautomatic pistol like the Glock 19 can be fired up to three times per second (0.333 s gap), so the minimum 0.973 gap in this recording is sufficiently long that a single firearm *could* account for all six shots.

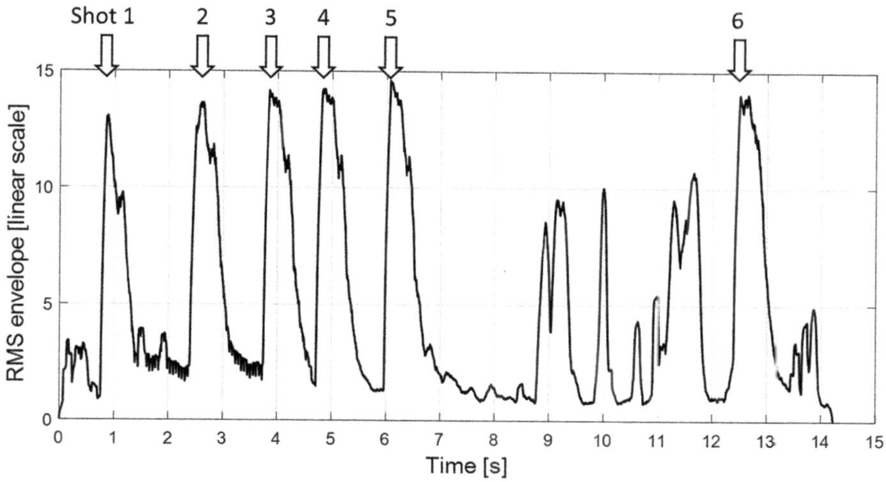

Fig. 9.20 Signal energy envelope (100 ms smoothing) for the signal of Fig. 9.18

Shot	Shot onset time (s)	Time gap between shots (s)
1	6.800	–
2	8.442	1.642
3	9.785	1.343
4	10.758	0.973
5	12.009	1.251
6	18.436	6.427

The signal energy envelope is calculated by squaring, smoothing, and then taking the square root of the signal. The envelope with a 100 ms smoothing is shown in Fig. 9.20.

As was evident in the critical listening step, the waveform envelope analysis shows general similarity between the shots, but there are differences in the overall level and detail of each one. Overlapping the six shots by aligning the onset times leads to the view shown in Fig. 9.21. Note the overall similarity, especially for shots 3–6.

One noticeable feature of the gunshot signal envelopes in Fig. 9.20 is the presence of a secondary pulse, appearing as a small "thumb" after the envelope peak of shots 1–5. Measuring the time difference between the onset of the shot envelope and the peak of the secondary pulse gives: Shot 1, 0.40 s; Shot 2, 0.47 s; Shot 3, 0.45 s; Shot 4, 0.45 s; and Shot 5, 0.45 s. The origin of this signal energy is not known. It could be a distinct acoustic echo from a large reflecting surface of some kind, or it could be some peculiarity of the audio recording device, such as an automatic gain control. If it was due to a reflection, the roughly 0.45-s delay at the speed of sound (c) would indicate the reflecting surface was 0.45/2 = 0.225 s away or approximately 77 m if c = 343 m/s. If this particular observation became important, the

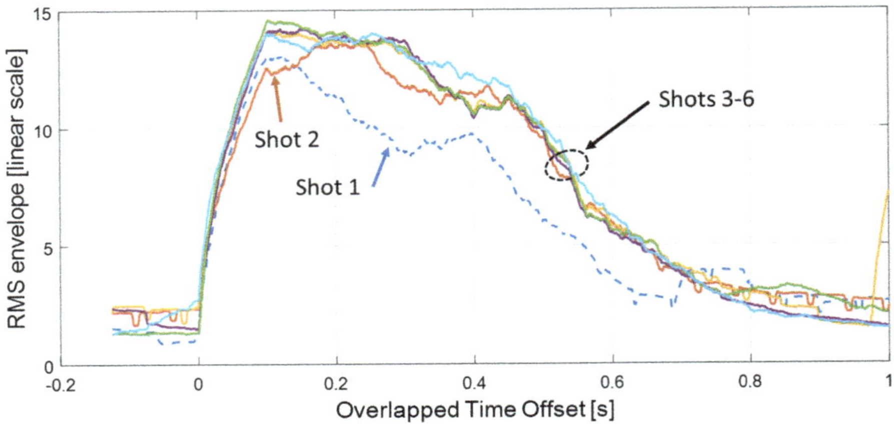

Fig. 9.21 Overlapped plot of the six shot envelopes

audio forensic examiner would need to know more details about the shooting scene and the ambient air temperature at the time of the incident to calculate the appropriate speed of sound—which is temperature dependent, as described previously.

Finally, from *spectral analysis*, the spectrogram confirms the overlap of the Taser pulses and gunshots, the presence of speech utterances, and the background noise level of the recording.

Thus, the results of critical listening, waveform analysis, and spectral analysis can provide information that addresses the question "*Are the gunshot sounds all attributable to a single firearm?*" as follows:

- The six shots have similar subjective sounds and similar waveform envelopes. The objective audio evidence also indicates noticeable similarity, including the overall sound level, reverberant decay, and secondary pulse/echo.
- The first two shots have a different amplitude envelope than the remaining four shots. This could indicate that the first shots were from a different firearm than the latter shots or that a single firearm was fired from a moving position. The small audio differences between each shot could be attributed to a single firearm moving or changing orientation between the shots, although the observations do not rule out the possibility that shots came from two similar firearms, all in proximity to each other.
- The inter-shot timing is at least 973 ms, which a firearm expert could verify as being sufficient for successive shots from the same semiautomatic pistol.

It is ultimately the job of the detectives and the attorneys to combine the uncertain audio forensic findings with other evidence and witness testimony in order to build their case or their defense.

Question 2: Is the Taser Deployment Before or After the First Gunshot?

As noted above, the Taser device, when deployed, may emit an audible clicking sound. The pulses occur with a 50 ms period and continue for up to 5 s. The portion

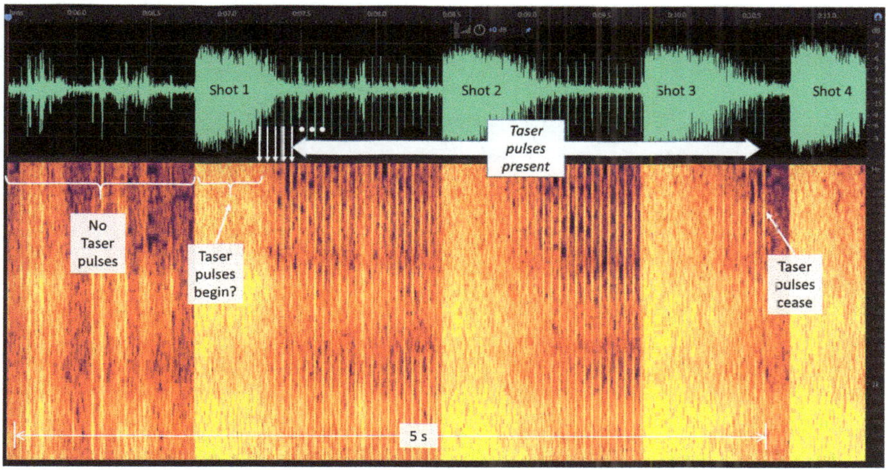

Fig. 9.22 Recording of overlapping sound of Taser device pulses and Shot 1 (overall duration 6 s, frequency range 400 Hz–3.1 kHz, linear scale)

of the recording with the overlapping sound of the first gunshot and the Taser device pulses is shown in Fig. 9.22.

The Taser pulses are clearly detected and visible in the time waveform and in the spectrogram starting as the sound of Shot 1 decays. The pulses cease before Shot 4, and no pulses are detected prior to Shot 1. The intense onset of Shot 1 overwhelms the recorded signal, so there is uncertainty about the exact moment the first Taser pulse occurs. All that can be stated is that the Taser pulses initiate at some point in the 0.4 s between the onset of Shot 1 and the clear appearance of the pulses in the waveform and spectrogram.

An expanded view of Shot 1 is shown in Fig. 9.23. The arrows identifying the distinct Taser pulse sequence visible in the right side of the figure has been extrapolated earlier in time with matching steps of 50 ms, to indicate where any prior pulses would be expected.

According to the Taser ECW literature, the device is expected to perform its pulsing discharge cycle for approximately 5 s after the trigger is pulled and released. However, in the example forensic recording shown in Fig. 9.22, a 5-s interval prior to the time at which the last Taser pulse is detected would indicate that the first pulse would presumably be prior to the onset of Shot 1. It cannot be determined from the audio evidence why the pulse sequence is audible for 3.4 s, not the expected 5 s. The case detective would need to consult an expert in the operation of the Taser device to help answer this question. Based on the audio evidence, it seems that the onset of Shot 1 occurs before, or possibly concurrently with, the Taser device deployment.

Question 3: Were the Speech Sounds Uttered by a Man or by a Woman?
In general, male talkers speaking at a normal conversational level are found to have fundamental frequencies (F_0) in the range 85–180 Hz, and female talkers speaking naturally have typical fundamental frequency range 165–255 Hz. It is important to

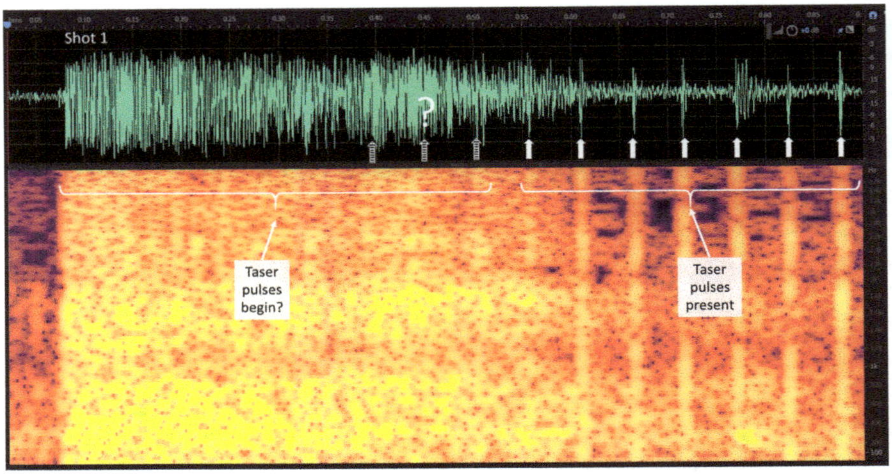

Fig. 9.23 Expanded view of the signal and spectrogram of Shot 1 (overall duration 0.9 s, frequency range 0–3.8 kHz, linear scale)

notice that the male and female frequency ranges actually overlap, so while the trend is that male voices have lower fundamental frequencies than female voices, this is only a general observation.

Critical listening to this example forensic recording (Fig. 9.18) provides the *subjective* impression that the utterances are *shouted* words by a male talker. One of the utterances from the forensic recording is shown in Fig. 9.24. The fundamental frequency of the shouting appears in the range 340–355 Hz.

We note in the forensic example recording that (a) the subjective impression is that a male talker is shouting and (b) the fundamental frequency is very high compared to the expected F_0 ranges for male and female talkers. Is this expected?

In the case of shouting, screaming, or other excited and emotional utterances, the fundamental frequency is typically much higher than for normal conversational speech. For example, an example waveform and spectrogram for a male talker saying the word "too" at a normal, comfortable speech level for a conversation is shown in Fig. 9.25. Note that the fundamental frequency is roughly 147 Hz, which is within the normal range of a male talker.

When the same male talker is recorded when being told to shout the word "too" at a very high level of vocal effort, as shown in Fig. 9.26, the spectrogram is significantly different. One of the effects of loud shouting is that the talker raises both the vocal amplitude and the fundamental frequency. In this case, the fundamental is approximately 340 Hz: much higher than the 147 Hz found for his speech at a normal conversational level.

Thus, the examination of the audio forensic example with critical listening and spectrographic analysis gives confidence that the shouted utterances in the forensic recording are from a male talker. As with the other hypothetical audio forensic questions, it is ultimately up to the detectives and the attorneys to combine audio forensic findings with the other available.

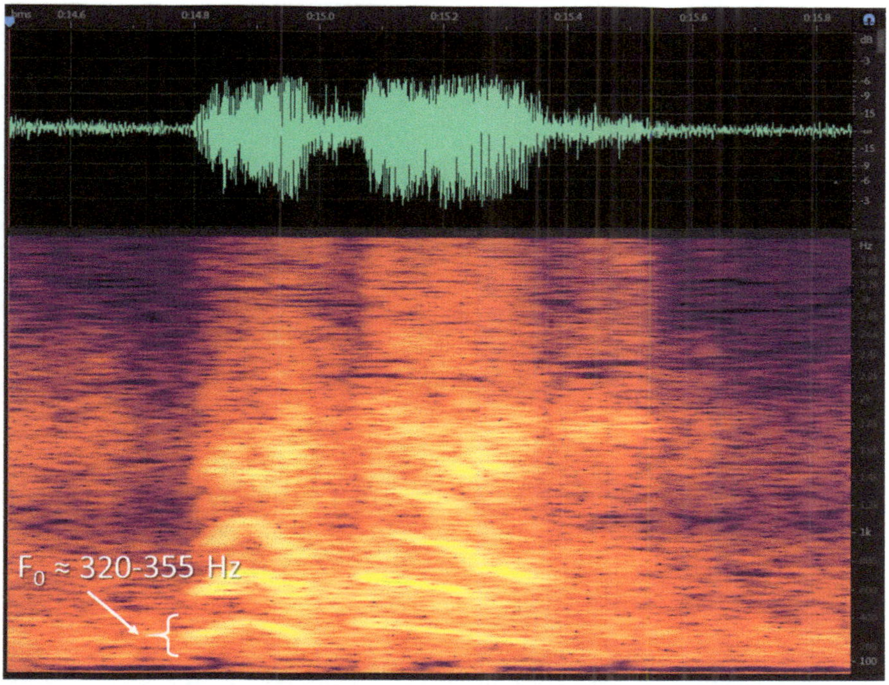

Fig. 9.24 Utterance from the forensic example recording (overall duration 1.35 s, frequency range 0–3.8 kHz, linear scale)

9.4.2 Example Forensic Recording 2: Gunshots with Multiple Recordings

As a second example of the type of gunshot recordings encountered by audio forensic examiners, Fig. 9.27 shows a recording of an incident involving multiple gunshots. The primary recording device, an officer's vest camera, was very close to where the firearms were discharged, resulting in a high degree of clipping and distortion.

Hypothetically, the detective investigating this case could pose a question for the audio forensic examiner, such as:

> Is it possible to estimate when the first few gunshots were fired?

Based on critical listening, waveform analysis, and spectral analysis of this vest camera recording, it is possible to identify at least four individual gunshots late in the barrage, as indicted in Fig. 9.27. Additional overlapping shots are audible in the 2–3 s prior to the first discernable gunshot, but the clipping and distortion is too extreme to resolve them.

Because the audio recording is of poor quality, the detective makes further inquiries and discovers that there were four law enforcement vehicles parked near the

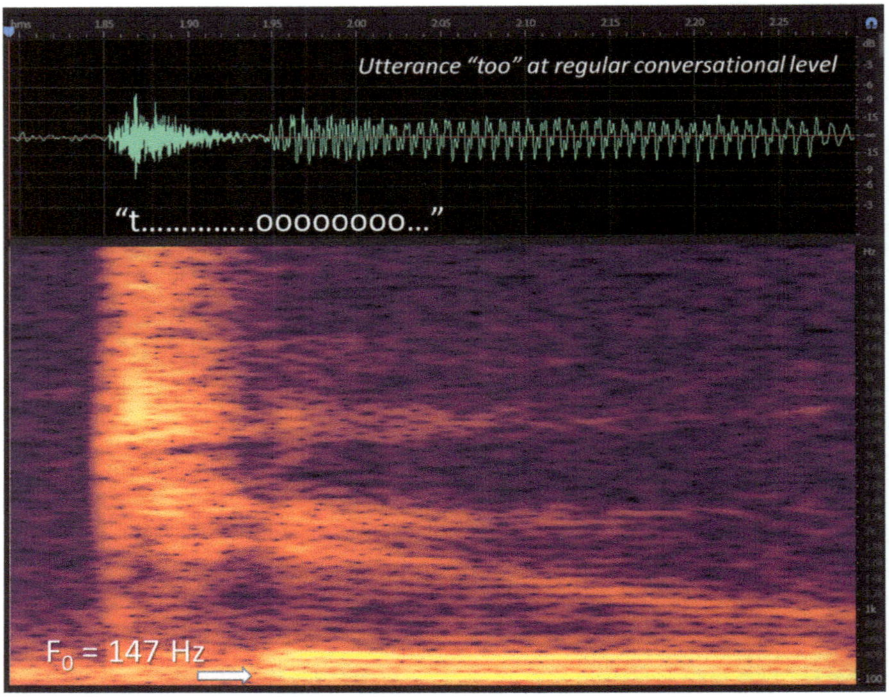

Fig. 9.25 A recording of a male talker uttering the word "too" at a normal conversational level of vocal effort (overall duration 0.5 s, frequency range 100 Hz–5.6 kHz, linear scale)

scene, and each vehicle was equipped with a dashboard audio/video camera system (a "dashcam"). The four vehicles were located 10 m or more from the shooting scene. Although the orientation of the dashboard cameras was such that none of the videos captured the shooting scene, the audio associated with the dashcam recordings was picked up by a microphone located in the cabin of each of the law enforcement vehicles. A collection of the four dashcam recordings and the original recording from the scene is shown in Fig. 9.28, with the unsynchronized recordings shifted to align the last audible shot in the sequence for each recording.

All five recordings cover the same time interval, but all the recordings were made with separate, unsynchronized devices. Comparing the five audio recordings, it is possible to identify uniquely the last gunshot, labeled "D," in each of the recordings and use that reference to align the five time bases. Of the five, the recording from Unit B1 provides the clearest view of the first four gunshots, labeled 1–4. The timing of each of the first four shots relative to Shot D can now be confirmed in recordings H4, B1, B4, and C1.

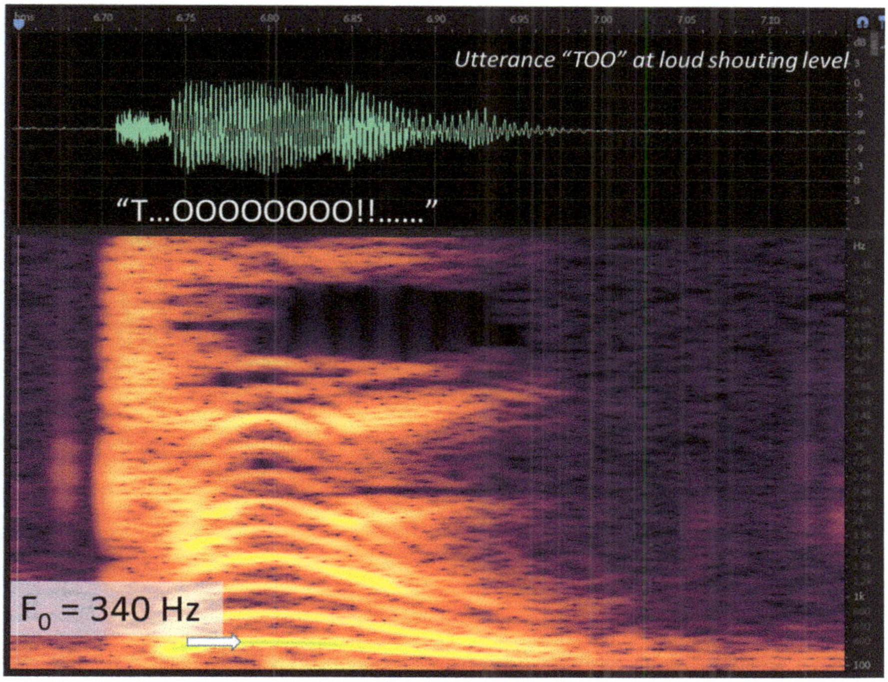

Fig. 9.26 A recording of a male talker shouting the word "too" with a high level of vocal effort (overall duration 0.5 s, frequency range 100 Hz–5.6 kHz, linear scale)

The recording from Unit B1 appears to have the least background noise and interference. Using that particular recording, the shot times expressed in seconds with respect to shot D are summarized:

	Shot 1	Shot 2	Shot 3	Shot 4	Shot A	Shot B	Shot C	Shot D
Unit B1	−4.329	−3.689	−3.461	−3.299	−1.257	−0.922	−0.508	0

Thus, it is possible to answer the audio forensic question regarding the timing of the first few shots in the sequence using a combination of the clipped and distorted vest camera recording, supplemented by the concurrent dashcam recordings.

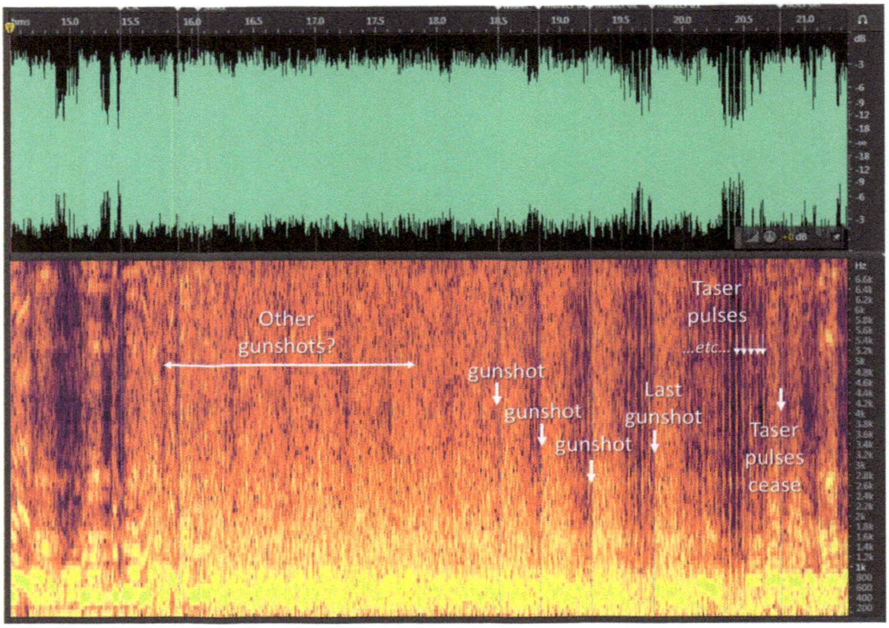

Fig. 9.27 Example forensic recording with multiple gunshots (highly overloaded and distorted due to close proximity) (overall duration 7 s, frequency range 0–6.8 kHz, linear scale)

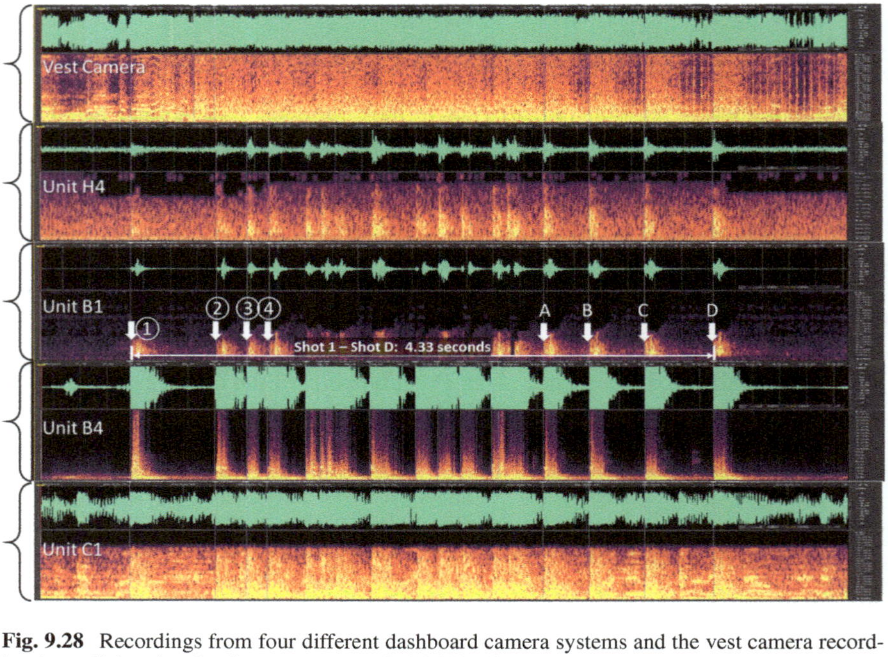

Fig. 9.28 Recordings from four different dashboard camera systems and the vest camera recording (overall duration 7 s, frequency range 0–6.8 kHz, linear scale)

References

Beck, S. D., Nakasone, H., & Marr, K. W. (2011). Variations in recorded acoustic gunshot waveforms generated by small firearms. *The Journal of the Acoustical Society of America, 129*, 1748–1759.

Brustad, B. M., & Freytag, J. C. (2005). A survey of audio forensic gunshot investigations. In *Proceedings of the Audio Engineering Society 26th Conference Audio Forensics in the Digital Age, Denver, CO*.

Koenig, B. E., Hoffman, S. M., Nakasone, H., & Beck, S. D. (1998). Signal convolution of recorded free-field gunshot sounds. *Journal of the Audio Engineering Society 46*(7/8), 634–653.

Maher, R. C. (2006). Modeling and signal processing of acoustic gunshot recordings. In *Proceedings of the IEEE Signal Processing Society 12th DSP Workshop, Jackson Lake, WY* (pp. 257–261).

Maher, R. C. (2007). Acoustical characterization of gunshots. In *Proceedings of the IEEE SAFE Workshop on Signal Processing Applications for Public Security and Forensics, Washington, DC*.

Maher, R. C. (2010). Overview of audio forensics. In *Intelligent multimedia analysis for security applications*. Berlin, Germany: Springer.

Maher, R. C., & Routh, T. K. (2017). Gunshot acoustics: Pistol vs. revolver. In *Proceedings of the 2017 Audio Engineering Society International Conference on Audio Forensics, Arlington, VA* (pp. 1–7).

Maher, R. C., & Shaw, S. R. (2008). Deciphering gunshot recordings. In *Proceedings of the Audio Engineering Society 33rd Conference Audio Forensics—Theory and Practice, Denver, CO* (pp. 1–8).

Routh, T. K., & Maher, R. C. (2016). Determining muzzle blast duration and acoustical energy of quasi-anechoic gunshot recordings. Preprint 9635. In *Proceedings of the 141st Audio Engineering Society Convention, Los Angeles, CA*.

Sadler, B. M., Pham, T., & Sadler, L. C. (1998). Optimal and wavelet-based shock wave detection and estimation. *The Journal of the Acoustical Society of America, 104*(2), 955–963.

Weissler, P. G., & Kobal, M. T. (1974). Noise of police firearms. *The Journal of the Acoustical Society of America, 56*(5), 1515–1522.

Chapter 10
Application Example 2: Cockpit Voice Recorders

Commercial passenger airline accidents may involve total destruction of the aircraft, damage to structures on the ground, and terrible loss of life. Air safety experts need to determine the cause of the accident in order to reduce the likelihood of similar accidents in the future, but when the aircraft is broken apart and burned, the cause of the accident may be a mystery. Fortunately, aviation regulatory agencies can afford to spend substantial resources to investigate the cause of accidents when they do occur, because commercial aircraft accidents are so remarkably infrequent.

Arguably, the most important development in accident investigations was the invention of the data recorder equipment required on all civilian commercial passenger flights, large private jets, and many military flights. In the United States and throughout most of the world, commercial passenger aircraft and many military aircraft are equipped with automatic *flight data recorder* (FDR) systems intended to survive a crash. A particularly interesting and unusual specialty in audio forensics is analysis of the audio recordings from *cockpit voice recorders* (CVRs). This audio forensic specialty is primarily reserved for individuals working for the US National Transportation Safety Board (NTSB), but private examiners employed by aircraft companies, airlines, and civil investigators may also be involved in analysis and interpretation of CVR data (National Transportation Safety Board 2007). An example of a CVR unit is shown in Fig. 10.1. Although press reports sometimes refer to data recorders as "black boxes," the devices are actually painted bright orange to make them easier to locate amid debris.

Developed in the 1950s and 1960s, flight data recorder devices keep a record of various flight parameters and record cockpit conversations and other sounds. Originally developed using special fireproof copper foil and later using fire-resistant magnetic tape, contemporary FDRs and CVRs now record in digital form using nonvolatile solid-state memory—although some older units in use may still contain magnetic tape. Older CVRs recorded a 30-min loop before overwriting, but current CVR devices record at least 2 h of audio.

R. C. Maher, *Principles of Forensic Audio Analysis*, Modern Acoustics and Signal Processing, https://doi.org/10.1007/978-3-319-99453-6_10

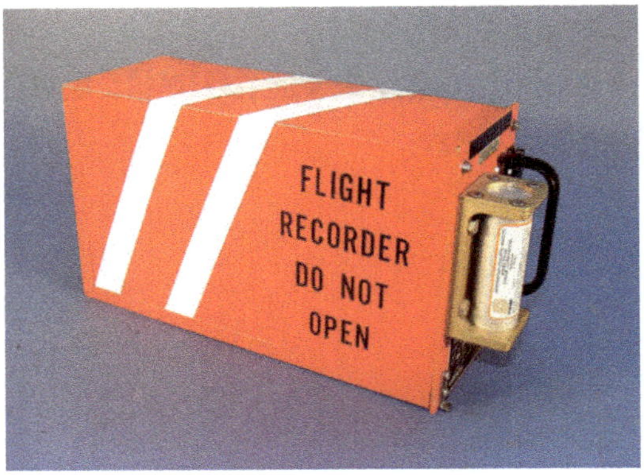

Fig. 10.1 Example of a cockpit voice recorder (CVR) chassis. The data recorders are painted bright orange to aid in finding the device in crash wreckage (NTSB 2015)

The FDR maintains a record of flight parameters such as time of day, altitude, aircraft orientation, airspeed, and so on. FDR systems on contemporary airliners can record hundreds of flight parameters, actuator positions, and sensor readouts every second, with memory capacity for up to 25 h (National Transportation Safety Board 2015).

Yet, even with the plethora of FDR digital information, the acoustical information from the CVR is often indispensable for accident investigators to piece together what happened leading up to the accident. In addition to providing information about conversations and telltale background sounds during flight, CVR systems are typically activated automatically and start recording whenever the aircraft is powered up, whereas the FDR systems collect flight data only from the point at which the plane begins accelerating down the runway to become airborne. This means that the CVR may contain important information about flight checklist completion, preflight discussion, and similar audio information obtained before takeoff that is not covered by the airborne FDR information.

When widespread use of cockpit voice recorders was first proposed in the 1960s, one of the biggest concerns was protecting the privacy of the flight crew and others whose utterances might be recorded during a flight. Some early CVR designs allowed the recorded audio to be erased at the conclusion of a safe flight. In the United States, the NTSB is given authority to investigate civil aircraft and surface transportation accidents, and by law the NTSB may not routinely release the CVR recordings and transcripts. The law does allow that "The Board shall make public any part of a transcript or any written depiction of visual information the Board decides is relevant to the accident or incident…" (United States Code 2009), but confidentiality and privacy remains the key consideration.

Besides the safety applications in aircraft cockpits, voice and data recorders are also used, or proposed for use, in other public transportation systems, such as the cab of a railroad locomotive, the bridge or wheelhouse of a ship or ferry vessel, and the interior of passenger buses.

10.1 CVR Operation and Interpretation

Current CVRs capture four separate monophonic channels: the pilot's headset microphone, the copilot's headset microphone, a cockpit area microphone (CAM) mounted in the cockpit's ceiling panel, and the fourth channel that may be used to record the intercom communications between the pilots and the flight attendants, the cabin public address system, or for some other purpose. Modern CVRs record up to 120 min of audio in a memory buffer loop, sequentially overwriting the oldest data with new data. Thus, in the event of an accident, the audio forensic examiner will have a recording containing the sounds from the 2 h preceding the crash (National Transportation Safety Board 2007).

NTSB audio forensic experts analyze the four audio channels for speech utterances and background sounds. Specialists transcribe the spoken words of the flight crew. The CAM system can also pick up non-speech sounds that are often very important to the investigation. Engine sounds, airframe vibrations, avionics audible warning alarms, and sounds of cockpit intrusions or other commotion may be detected.

For example, in March 2015 there was a crash of an Airbus 320 aircraft in the French Alps, killing all 150 passengers and crew on board the plane. The seemingly routine journey of Germanwings Flight 9525 took place in good weather. About 30 min into the flight, radar tracking indicated that the aircraft had started descending steadily, apparently under control, but without air traffic control authorization. The plane did not respond to radio calls from air traffic control. Ten minutes later, the plane collided with remote mountains in southeast France, and there were no survivors. There was no communication with ground controllers, no witness explanations, and no remaining physical evidence on the ground after the crash to help investigators understand what happened.

Instead, the tragic demise of Germanwings Flight 9525 was soon explained due to evaluation of the cockpit voice recorder evidence. Investigators from the French and German civil aviation safety authorities recovered the damaged but usable CVR and discovered that when the pilot left the cockpit momentarily to use the lavatory, the copilot had apparently locked the cockpit door deliberately to prevent the pilot from reentering the flight deck. The copilot, alone in the cockpit, then took steps to crash the aircraft intentionally. The investigation report describes audio from the CVR, including the urgent pleas from the pilot, the sound of futile attempts to break down the cockpit door, and even the recorded sound of the steady breathing of the copilot, indicating that he was not incapacitated. There was no mechanical issue with the airplane: It was a suicidal copilot. Following the tragic incident, many airlines adopted the policy that the flight deck must have two individuals present at all times to reduce the likelihood of unilateral action by a distraught individual.

Later in 2015, a British Aerospace Hawker 700 aircraft operating as Execuflight Flight 1526 crashed on approach for landing at Akron, Ohio, on November 10, 2015. The crash investigators wanted to understand the jet engine throttle settings and turbine speed parameters, but the engine control system had been destroyed by the crash and ensuing fire, and the particular aircraft was old enough that no Flight Data Recorder (FDR) was required. The plane did have a CVR, but it was an older model in poor condition with a 30-min loop of analog 4-channel audio tape. Nonetheless, the investigators were able to use the CVR recordings to identify the whine of the engine turbines. They attempted to deduce the various jet engine operational parameters from the audio frequencies of the engine noise. The NTSB conclusions about the crash placed responsibility on the captain and first officer for not following the proper approach checklist and for not aborting the landing after recognizing that the aircraft speed and rate of descent were likely to cause an aerodynamic stall (NTSB 2016).

A 1997 investigation of the CVR data from a Beechcraft 1900C commuter aircraft accident that occurred in 1991 used signal characteristics from both the cabin microphone and an unused CVR channel. The researchers were studying a theory that an in-flight engine separation was preceded by evidence of propeller whirl flutter attributable to a cracked truss in the engine mount (Stearman et al. 1997). This case is interesting because a prior official NTSB report of the accident includes examination of the CVR information and transcript, but does not include observation of the background sounds in the recording (NTSB 1993). This type of technical dispute indicates the importance of having multiple sources of flight data information and expertise involved in the investigation.

In another significant prior case involving audio forensic investigation using CVR data, examiners from the NTSB focused on audio recordings from the September 1994 crash near Pittsburgh of USAir Flight 427 (a Boeing 737 aircraft). The audio examiners sought to understand the behavior of the aircraft's engines and the timing, reactions, and efforts of the pilot and first officer during the incident. The CVR captured the rapid transition from routine cockpit activity and radio communication to the abrupt onset of an in-flight emergency. NTSB investigators were able to assess the pilots' effort, state of awareness, rate of respirations, and other important aspects of the emergency response. Among other clues, several audible clunks and rattles in the CVR recordings led the investigators to try several experiments to determine the ability of the cockpit microphone to pick up sound through structure-borne vibration (NTSB 1999).

The circumstances and post-accident evidence from USAir 427 led investigators to consider the similarity with one prior and one subsequent unexplained incidents involving Boeing 737 aircraft: United Airlines Flight 585 (March 3, 1991) and Eastwind Airlines Flight 517 (June 9, 1996). The prior United Airlines crash occurred on approach to Colorado Springs, killing all those aboard the aircraft, while the Eastwind Airlines incident a few years after USAir 427 involved a near-loss of control of the aircraft, but the pilot was able to recover control and the aircraft landed without injury. Ultimately, after years of investigation, the NTSB determined that a defect in the operation of a hydraulic device known as the rudder power control unit, or rudder PCU, was likely the cause of the three incidents. The PCU defect resulted in a sudden *rudder reversal*: Instead of moving in the direction commanded by the pilot's foot pedals, the vertically mounted rudder surface on the tail moved to the extreme opposite position. This defect was repaired in newly designed rudder PCU units installed on all 737 aircraft (NTSB 1999; Byrne 2002).

10.2 The Future Role of Audio Forensics in Transportation Safety Systems

Like most areas of digital technology, future flight data and cockpit voice recorders will incorporate many advanced features and capabilities. Among the current concerns have been recent airliner crashes in which contact with the aircraft was lost over the open ocean and the wreckage containing the data recorders was extremely difficult or impossible to locate.

For example, the loss of Air France Flight 447, an Airbus A330 aircraft, took place over a remote area of the Atlantic Ocean between South America and Africa on June 1, 2009. Some floating wreckage was found within the first few days after the crash, but the portion of the aircraft containing the FDR and CVR sank in waters approximately 3000 m (9800 ft.) deep. The wreckage on the sea bottom was not located until May 2011, nearly 2 years after the accident. The FDR and CVR were recovered and provided key information in understanding the cause of the accident, but the substantial delay between the accident and the data recovery left potential risks unsolved for years (Bureau d'Enquêtes et d'Analyses 2012).

Another example is Malaysia Airlines 370, a Boeing 777 aircraft that disappeared from radar during what appeared to be a routine flight from Kuala Lumpur on March 8, 2014. As of the time of this writing (July 2018), no major concentration of wreckage has been found anywhere in the vast search area west of Australia in the Indian Ocean. It is not known if the FDR and CVR will ever be found, leaving an unsolved mystery for aviation safety (Encyclopaedia Britannica 2018).

Because of these and other examples, aircraft engineers and accident investigators are calling for systems that will make the data recorders more easily recovered. Suggestions include developing mechanisms that automatically eject the data recorders from a crashing aircraft for easier recovery, or development of advanced radio beacon systems for all aircraft that would continuously stream the flight information wirelessly to orbiting satellites, thereby eliminating the need to recover the recorders after a crash.

References

Alexander, A., Forth, O., & Tunstall, D. (2012). Music and noise fingerprinting and reference cancellation applied to forensic audio enhancement. In *Proceedings of the Audio Engineering Society 46th International Conference Audio Forensics, Denver, CO.*

Begault, D. R., Heise, H. D., & Peltier, C. A. (2014). Forensic musicology: An overview. In *Proceedings of the Audio Engineering Society 54th International Conference Audio Forensics, London, UK.*

Bureau d'Enquêtes et d'Analyses. (2012). *Final report on the accident on 1st June 2009 to the airbus A330-203 registered F-GZCP, operated by Air France, flight AF 447, Rio de Janeiro – Paris*. Le Bourget, France: French Civil Aviation Safety Investigation Authority.

Byrne, G. (2002). *Flight 427: Anatomy of an air disaster*. New York: Springer.

Encyclopaedia Britannica. (2018). *Malaysia airlines flight 370 disappearance*. Retrieved from https://www.britannica.com/event/Malaysia-Airlines-flight-370-disappearance

Moore, A. H., Brookes, M., & Naylor, P. A. (2014). Room identification using roomprints. In *Proceedings of the Audio Engineering Society 54th International Conference Audio Forensics, London, UK*.

National Transportation Safety Board. (1993). *Loss of Control, Business Express, Inc., Beechcraft 1900C N811BE, Near Block Island, Rhode Island, December 28, 1991*, Aircraft Accident Report NTSB/AAR-93/01/SUM. Washington, DC.

National Transportation Safety Board. (1999). *Uncontrolled Descent and Collision with Terrain, USAir Flight 427, Boeing 737-300, N513AU, Near Aliquippa, PA, September 8, 1994*, Aircraft Accident Report NTSB/AAR-99/01. Washington, DC.

National Transportation Safety Board. (2007). *Cockpit voice recorder handbook for aviation accident investigations*.

National Transportation Safety Board. (2015). *Cockpit voice recorders and flight data recorders*.

National Transportation Safety Board. (2016). *Crash During Nonprecision Instrument Approach to Landing, Execuflight Flight 1526, British Aerospace HS 125-700A, N237WR, Akron, Ohio, November 10, 2015*, Aircraft Accident Report NTSB/AAR-16/03. Washington, DC.

Sachs, J. S. (2003). Graphing the voice of terror. *Popular Science, 262*(3), 38–43.

Stearman, R. O., Schulze, G. H., & Rohre, S. M. (1997). Aircraft damage detection from acoustic and noise impressed signals found by a cockpit voice recorder. *Proceedings of the National Conference on Noise Control Engineering, 1*, 513–518.

United States Code. (2009). *49 U.S.C. 1114 – Disclosure*. Availability, and use of information. U.S. Government Publishing Office. Retrieved from https://www.gpo.gov/fdsys/pkg/USCODE-2009-title49/pdf/USCODE-2009-title49-subtitleII-chap11-subchapII-sec1114.pdf

Chapter 11
Conclusion

This book was written to emphasize the basic principles of audio forensic analysis and to serve as a primer for readers interested in developing expertise in one or more facets of audio forensics. While analog tape recorders dominated the field for many years, contemporary work is nearly always based upon digital recordings and digital signal processing. This fact means that audio analysis may be only a small part of a larger forensic investigation involving digital video, still pictures, digital file recovery, modification logs, encryption, and many other areas of digital computer forensics. Digital audio forensics experts increasingly need to develop skills in handling digital media of all kinds.

Ongoing challenges for audio forensic analysis have to do with the increasing use of lossy perceptual audio coding, such as MPEG. The lossy audio coding provides excellent perceptual quality for human listeners at low bit rates, but the effects of the data compression may make waveform interpretation more complicated than for uncompressed PCM recordings. Future audio formats and privacy encryption will undoubtedly continue this trend for encoded perceptually compressed information, and this trend will continue to complicate attempts to authenticate and interpret digital audio recordings.

On the other hand, technological trends indicate the emergence of new opportunities as more forensic evidence becomes available due to the rapid increase in the number of law enforcement agencies that require officers to use vest cameras, dashboard cameras, and facility surveillance systems. Similarly, many businesses and even regular citizens are now installing private security camera systems, which may also record audio. Moreover, the incredible growth in the number of people who carry mobile phones with audiovisual capability also means a growing likelihood that recordings capturing significant events and incidents will be available for forensic analysis. In summary, it seems clear that the growing number and increasing quality of audio recording devices will provide a substantial supply of audio forensic material for many investigations.

R. C. Maher, *Principles of Forensic Audio Analysis*, Modern Acoustics and Signal Processing, https://doi.org/10.1007/978-3-319-99453-6_11

New areas of research continue to emerge in the audio forensics field:

- Acoustical modeling, sometimes referred to as "acoustical fingerprinting," has been proposed as a means to authenticate the location a forensic recording took place and, possibly, to detect the particular equipment used to make the recording (Alexander et al. 2012; Moore et al. 2014).
- There are many interesting research leads involving computer-assisted classification of speech, pattern detection and recognition, and forensic talker identification (Sachs 2003; Begault et al. 2014).
- Acoustical beamforming and multilateration techniques are being developed to take advantage of situations in which multiple microphones are present at the scene of an incident. The multiple channels can conceivably allow better localization of particular sound sources, increased signal-to-noise ratio, or adaptive noise reduction.
- Techniques developed for post-incident forensic study of recordings may also be applicable for real-time detection and processing "on the fly" (tactical use) as faster and more capable signal processing systems become available.

References

Alexander, A., Forth, O., & Tunstall, D. (2012). Music and noise fingerprinting and reference cancellation applied to forensic audio enhancement. In *Proceedings of the Audio Engineering Society 46th International Conference Audio Forensics, Denver, CO*.

Begault, D. R., Heise, H. D., & Peltier, C. A. (2014). Forensic musicology: An overview. In *Proceedings of the Audio Engineering Society 54th International Conference Audio Forensics, London, UK*.

Moore, A. H., Brookes, M., & Naylor, P. A. (2014). Room identification using roomprints. In *Proceedings of the Audio Engineering Society 54th International Conference Audio Forensics, London, UK*.

Sachs, J. S. (2003). Graphing the Voice of Terror. *Popular Science, 262*(3), 38–43.

Index

R. C. Maher, *Principles of Forensic Audio Analysis*, Modern Acoustics and Signal Processing, https://doi.org/10.1007/978-3-319-99453-6

www.ingramcontent.com/pod-product-compliance
Ingram Content Group UK Ltd.
Pitfield, Milton Keynes, MK11 3LW, UK
UKHW021410091026
15624UKWH00009B/94
* 9 7 8 3 3 1 9 9 9 4 5 2 9 *